T0241029

Offstage Space, Narrative, and the Theatre of the Imagination

Offstage Space, Narrative, and the Theatre of the Imagination

William Gruber

OFFSTAGE SPACE, NARRATIVE, AND THE THEATRE OF THE IMAGINATION
Copyright © William Gruber, 2010.
Softcover reprint of the hardcover 1st edition 2010 978-0-230-62289-0

All rights reserved.

First published in 2010 by
PALGRAVE MACMILLAN®
in the United States—a division of St. Martin's Press LLC,
175 Fifth Avenue, New York, NY 10010.

Where this book is distributed in the UK, Europe and the rest of the world,
this is by Palgrave Macmillan, a division of Macmillan Publishers Limited,
registered in England, company number 785998, of Houndmills,
Basingstoke, Hampshire RG21 6XS.

Palgrave Macmillan is the global academic imprint of the above companies
and has companies and representatives throughout the world.

Palgrave® and Macmillan® are registered trademarks in the United States,
the United Kingdom, Europe and other countries.

ISBN 978-1-349-38449-5   ISBN 978-0-230-10564-5 (eBook)
DOI 10.1057/9780230105645
Library of Congress Cataloging-in-Publication Data

Gruber, William E.
    Offstage space, narrative, and the theatre of the
    imagination / William Gruber.
        p. cm.

    1. Offstage action (Drama) 2. Drama—Technique.
    3. Narration (Rhetoric) I. Title.
PN1696.G78 2010
808.2—dc22   2009030402

A catalogue record of the book is available from the British Library.

Design by Newgen Imaging Systems (P) Ltd., Chennai, India.

First edition: March 2010

Transferred to Digital Printing in 2011

This book is for Robert

CONTENTS

Introduction: Sights Unseen

I have observed that in all our tragedies, the audience cannot forbear laughing when the actors are to die; it is the most comic part of the whole play.
—John Dryden, *Essay of Dramatic Poesy* (1668)

Near the end of *The Winter's Tale*, after Leontes has been reunited with his daughter, Perdita, and with Polixenes, the King of Bohemia, three characters identified only as "gentlemen" come on stage to give a retrospective account of that joyful meeting. Superficially the scene is a thinly dramatized narrative of events which have taken place recently off stage; many find it odd that Shakespeare chose to tell about the reunion rather than show it, and for this reason the scene has not been much admired. Samuel Johnson, for example, attributed its recursive format to slothfulness on the part of its author: "It was, I suppose, only to spare his own labour that the poet put this whole scene into narrative, for though part of the transaction was already known to the audience, and therefore could not properly be shewn again, yet the two kings might have met upon the stage... and the young lady might have been recognized in sight of the spectators." Here is part of the scene that caused Johnson to doubt Shakespeare's energy:

> 3. Gentleman. Did you see the meeting of the two kings?
> 2. Gentleman. No.
> 3. Gentleman. Then you have lost a sight which was to be seen, cannot be spoken of there might you have beheld one joy crown another, so and in such manner that it seemed sorrow wept to take leave of them, for their joy waded in tears. There was casting up of eyes, holding up of hands, with countenance of such distraction that they were to be known by garment, not by favor. Our king, being ready to leap out of himself for joy of his found daughter, as if that joy were now become a loss, cries, "O, thy mother, thy mother!" then asks Bohemia forgiveness;

then embraces his son-in-law; then again worries he
his daughter with clipping her; now he thanks the old
shepherd, which stands by like a weather-bitten conduit
of many kings' reigns. I never heard of such another
encounter, which lames report to follow it and undoes
description to do it. (5.2.40–57)

Johnson is a shrewd critic of theatre, and nothing in what he says
about Shakespeare's dramaturgy can be attributed to ignorance or
wrong-headedness with respect to the dependence of effective theatre
on scenic enactment. Most of us would agree with the general proposition that underlies Johnson's complaint, namely, that in comparison
with actually seeing something, words are usually a poor substitute. For
modern theatre audiences this mindset is especially uncompromising;
it has been so deeply fixed for contemporary audiences by widespread
exposure to visual media that nowadays even those things which are entirely accessible in and through language are invariably given a pictorial
dimension. One cannot, for example, listen to television broadcasters
report the comparative statistics of holiday sales figures without at the
same time seeing an accompanying sequence of images of debit cards,
cash registers, and pairs of hands counting out piles of greenbacks.

Viewed, however, from a different, more indulgent angle, this particular scene from *The Winter's Tale* appears in some respects to be
quintessentially Shakespearean. The comments of the various gentlemen regarding the relative inefficacy of narrative in representing events
are surely meant to invite spectators to think of their own role as both
viewers *and* listeners and to consider the scene with a degree of aesthetic
detachment. Moments such as this when Shakespeare turns his mirror
onto his own art can be found throughout his work from its beginnings.
But especially in late dramas such as *The Tempest* or *The Winter's Tale*,
Shakespeare more than ever before seems to be amused by the conventions of theatre on which he has relied for so long, making light of the
features of his art at the same time he depends on them. Such seems
to be the case with the foregoing scene from *The Winter's Tale*: one
character (Gentleman 3) tells the audience of an event so marvelous it
cannot be described—at which point he flatly contradicts himself, pro-
ceeding to paint with words a rich and lively image of the meeting he's
just claimed was beyond words' reach. The artistic sleight-of-hand by
which narrative replaces enactment is so cleverly ironic one can imagine
Shakespeare composing the scene on a wager: a master playwright bets
that he can first cause his audience to believe that they missed out on

seeing something important and then provides them with a fully imaginary verbal account that attempts to match—if not in some ways to surpass—the vitality and vivacity of any actual dramatization.

"Each of the arts," writes Elaine Scarry, "incites us to the practice of…immediate perception, delayed perception, and mimetic perception. But painting, sculpture, music, film, and theater are weighted toward the first…whereas the verbal arts take place almost exclusively in the third" (Scarry, *Dreaming* 7). Scarry's analysis of the different kinds of aesthetic responsiveness to art offers a way to parse further Johnson's complaint about Shakespeare's dramaturgy. Even though Johnson is, in fact, commenting on his reading of the play, he adopts the point of view of a spectator to a theatrical performance. He tends to judge Shakespeare's art, therefore, from the attitude of one who expects a multidimensional experience consistent with what Scarry calls "immediate perception," and it is this expectation of sensory perception that causes him to feel cheated when he is forced to see events with his imagination rather than with his eyes.

Johnson's experience is one known to everybody who has ever been disappointed by the inadequacy of a daydreamed face or by a verbal account of a beloved landscape. We tend especially to notice the relative poverty of words as a stand-in for sight when for some reason we are particularly keen on seeing something specific or when we want to attend to an event with as much clarity and knowledge as is possible. It is exactly this kind of satisfaction that comes when our expectations are met that Johnson claims is lacking in *The Winter's Tale*. With regard to the gentlemen's conversation, upon learning that a crucial recognition scene has been passed over, we are alerted to the discrepancy between seeing something when it happens and learning about it second hand, and it is under circumstances such as these that we are likely to feel that in being merely told about an event we have been given something deficient.

The reason for Johnson's disappointment is not that he believes that narrative on stage is necessarily an inadequate mode of representing human actions, in and of itself; as someone who fully admires the comparatively restrictive dramaturgy of the classical tradition, Johnson surely understands that some classes of events are better said than seen. The problem, rather, is that Johnson (and in this he speaks for many other readers) feels victimized by what appears to be a strategy of bait-and-switch: most spectators are likely to be disappointed when they are given words in a situation where they had been prepared to want a theatrical experience with more immediate sensory content. For Johnson, this substitution amounts to something like a playwright's dereliction of duty; hence he imagines Shakespeare to be lazy and criticizes him for wanting spectators to accept a "mimetic" perception when they had been led—because they are watching a sequence of actions that seems to have been building to a "recognition scene"—to expect immediate perception.

But shadowing Shakespeare's curious refusal to make a scene in *The Winter's Tale* where a scene seemed called for is a related and more subtle problem of dramatic composition. The gentlemen's conversation represents a kind of implicit consideration of the boundaries of drama, in that it calls attention to the fact that the playwright here attempts to generate a vivid image of an action only by recounting it in words; Shakespeare thwarts the expectations of the audience by giving them a retrospective account of the reunion rather than by materially representing it. The scene thus exposes with peculiar force two questions regarding dramatic composition that are central to every playwright: (1) How to choose which events of a life to dramatize on stage, and which to express only in retrospect, by recitation? (2) How to establish and maintain a productive relationship between enactment and narrative in theatre?

The classic answer to the question of what the dramatist should show, on the one hand, and what should merely be told about, on the other, is given by Racine in the preface to *Britannicus*: "one of the rules of theater is to make a narrative only of those things that cannot occur in action." In thus setting limits on those actions and events that a playwright ought to dramatize, Racine's object is to install the concept of verisimilitude as the basis for a coherent dramaturgical practice. In selecting the things to show on stage, the task of the dramatist, according to Racine, is to strike a balance between such kinds of characters, actions, or objects that spectators would be likely to accept exclusively in terms of their values as signs, and those kinds that cannot easily be assimilated into the signifying system. For example, certain objects are too unyieldingly "real" or "raw" for the stage (a functioning clock on the wall is a famous example), while some actions, if they are simulated (such as an actor's pretending to die), appear too overtly "theatrical." According to Racine, these latter elements of any theatrical representation (e.g., the passage of extended lengths of time, deaths, or events that might be considered either marvelous or fantastic) are to be assigned to the offstage, whence they can be made accessible by means of linguistic report.

Racine makes it sound as if these same dramaturgical principles apply in all places and at all times, but his neoclassical bias with respect to the definition of verisimilitude is readily apparent. Most recent commentators, in contrast, tie any attempt to produce the illusion of truth in the theatre to the values, ideologies, and rhetorical codes of a particular historical era. Of course, these artistic codes vary widely, especially as regards the staging of scenes involving physical violence or dying. Quite apart from questions as to the mimetic capabilities of actors or the relative naiveté of audiences, for example, it is certainly the case that dramatists known for their willingness to stage violence and death—the Jacobean playwrights of the early seventeenth century would be one familiar example—could not have succeeded in the absence of a community of spectators who were willing to credit even fairly crude representations of dying with something of the import and actuality of real life. Assuming that contemporary accounts of typical Jacobean performances are accurate in their description of the actors' exaggerated histrionics when attempting to depict characters in extremis, one has to imagine an audience with a rather generous capacity to suspend disbelief, a generosity that is apparently not the case by the time Dryden (whose commentary I cite in the epigraph) noticed how funny it seemed when tragic actors pretended to die.

But Shakespeare's task in *The Winter's Tale* is not to dramatize the clash of large armies, multiple deaths, or sexual activity; his task is simply to make credible the meeting of family members who have long been forced to live apart from one another. Especially in this case, therefore, Racine's answer—to "make a narrative only of those things which cannot occur in action"— is no answer at all. When Shakespeare in the foregoing scene from *The Winter's Tale* deliberately bypasses enactment, what is actually laid bare is the question of whether the frustrations involved in listening to a relatively cumbersome narrative account of events can ever be made ultimately to serve "a more involved kind of satisfaction" (Burke 31). Here it could even be said that because the choice Shakespeare rejects—scenic enactment—seems so clearly to be the right one in this case, his abandonment of it, from a dramaturgical point of view, becomes all the more intriguing. For what reasons might Shakespeare want to trade the theatre's solid foundation in immediate perception, for the belated and apparently enfeebled products of the imagination?

In watching Hamlet speak a soliloquy, kill Polonius, or fence with Laertes, we engage in a direct mimesis (or "immediate perception," in

Scarry's terms) of seeing an actor present a character as living and moving before us; in contrast, in hearing Hamlet report on how he altered the letter belonging to Rosencrantz and Guildenstern on board the ship bound for England, or in listening to the third Gentleman's account of the meeting of the two kings, we shift the grounds of imitation from the stage to the imagination. The mimesis in this latter case is not so much "in" the staged representation as in our mental picturing of it, a picturing dependent almost entirely on the verbal template supplied by a retrospective narrative account. And the question, then, is this: how does this secondary kind of representation, a verbal construction that is commonly taken to be feeble and second-rate in comparison with the immediate perception of direct scenic enactment, come, in the right circumstances and under the control of a skilled playwright, to function as an adequate or possibly superior substitute for immediate visual phenomena? It is this less familiar side of dramatists' work of picking and choosing—the actions that playwrights elect not to show but to represent only in words—that will be the focus of this book.

More formally, in broadest terms, the book is concerned with *mimesis* and *diegesis* as they relate to the structure and performance of drama. I want to examine the relationship between these two modes of representation in the theatre, between direct scenic enactment, on the one hand, and the presentation of absent events through narration, on the other. I'm especially concerned to demonstrate the dramaturgically productive tension that has always existed between narration and enactment as modes of image-making in the theatre and to show how dramatists configure unseen people and events in ways that tap spectators' imaginations. For instance, one might ask, what is it that is lost—or gained—by having a messenger tell how Medea slaughtered her victims, describing at great length a sequence of ghastly occurrences that we ourselves cannot see or remember? Or one might ask of Terence's play *The Woman from Andros* how the eponymous main character (who never actually appears on stage) comes to possess dramaturgic weight? Or (to give one more introductory example), why do twentieth-century playwrights as different in their art and their politics as Yeats and Brecht almost always favor diegetic theatrical styles over more conventional mimetic forms, regularly framing their imitations of reality with the simultaneous autoreferentiality conferred by narrative?

I'm interested also to question the common assumption that dramatists choose "telling" over "showing" exclusively for reasons of expediency or taste. For instance, it's often assumed that in writing *The*

Persians it was dramaturgically easier for Aeschylus to create a single character who could report on the Battle of Salamis than to dramatize a naval combat involving in actuality more than a thousand triremes, or that in depicting the murder of a king in *Agamemnon* it was more tactful for him to have Clytemnestra describe her husband's assassination than to put the mayhem on public display. This may be so; Greek playwrights seem to have been unwilling to enact battle scenes in full public view, in contrast to Shakespeare, who seems to have gone out of his way to include them. But if we then go on to say that in representing murder and warfare Aeschylus was doubly constrained by the limits of technology and propriety and, therefore, was all but compelled to choose narrative as a default strategy, it tends to impede any further inquiry into the ways in which mental image-making on the part of the audience—as distinct from scenic enactment by the figures on stage—constitutes a functional and important part of classical tragic dramaturgy. In studying those components of drama that are seen only in the mind, in other words, I assume that technique in art is never just technique: it carries, as Geoffrey Hartman once observed, a responsibility toward the represented subject. Writing and performing the "unseen" can be every bit as dramaturgically complex as enactment. By asking when and under what circumstances a dramatist selects "telling" over "showing," then, I am asking about the reasons playwrights choose to substitute imagined events for perceptual ones, which is to say I am in the end asking about the role of the imagination in dramatic performance.

In order to clarify the kinds of "absences" I want to examine, let me digress briefly to explain that I'm not generally concerned with characters' passing remarks about things that happened yesterday, last month, or twenty years before the time of the action represented on stage. In conversations that take place in real life, people normally tell stories about the past—they tell their friends about a fight they had with their boss, about the snowfall last week, about a visit to the dentist. The same is true of plays, where characters often refer in retrospect to places, people, or events that the audience does not actually see dramatized. For example, during the opening moments of *The Cherry Orchard*, Lopakhin recollects his childhood: "I remember when I was a youngster about fifteen, my father—he's dead now but at the time he was a shopkeeper in the village here—hit me in the face with his fist. The blood ran out of my nose" (Chekhov 166). Lopakhin's references to the past go mainly to the development of his character. Reflections such as these are typically part of the norms of a realist dramaturgy; they are understood by audiences as having taken place "offstage" in what has been called "referred space" (McAuley 20) or "conceived space" (Scolnicov 14). These are places in which actions that form part of the structure of events of a play may be understood to have occurred, despite the fact that they do not contain actors and objects; even though such spaces are not shown, they are nevertheless imagined as physically locatable somewhere offstage. These locations differ from the "scenic" space on stage or the "perceived space" just off it (for example, a room or walkway that may be imagined to be located just outside the door on a set) chiefly in that they are not generally conceptualized in any explicit way by the performance or the mise-en-scène. Michael Issacharoff identifies this imaginary spatial dimension of almost any theatrical text as the diegetic space of the play, distinguishing it from mimetic (or shown) space because the latter "does not require mediation... [whereas] diegetic space is mediated by verbal signs (the dialogue) communicated verbally and not visually" (Issacharoff 55–56).

It follows from these analyses that the geography of Chekhov's play depends in part on the ability of an audience to imagine his characters as being embodied in these kinds of imaginary or "conceived" spaces. The many, apparently incidental remembrances of different times and different places give Lopakhin a greater degree of psychological complexity (or "roundness," to use E. M. Forster's well-known term), an illusion of solidity and depth that arises (just as in realist painting) from the conjunction of highly detailed but casually grouped narrative elements. The most striking image—that of the bloody nose—has no particular reason for being so prominent a feature of Lopakhin's memory, given the tender, nostalgic tone of his rambling. But the very randomness and surprising brutality of the detail—the blow from the beloved father, the bloody red splash on the face—is what gives the image its evocative power.

The Nurse's speech in *Romeo and Juliet* is another such instance:

> I remember it well.
> 'Tis since the earthquake now eleven years;
> And she was weaned (I never shall forget it),
> Of all the days of the year, upon that day,
> For I had then laid wormwood to my dug,
> Sitting in the sun under the dovehouse wall.
> My lord and you were then at Mantua. (1.2.23–29)

In referring briefly to previous events that the audience cannot see and cannot fully know, Shakespeare's and Chekhov's characters momentarily create for themselves a historical background, a depth-of-field against which they can be glimpsed as actors or agents. The technique endows them with the illusion of both solidity and temporal duration; because the information about their previous lives is only partial, it invites us to project an image of them existing and acting beyond our immediate perception. Ironically, these scattered fragments are far more successful at creating the illusion of "roundness" in these minor characters than a more formal biographical account. By means of such fleeting and incomplete temporal indexings, it is possible to infer a sense of an individual subject's duration, or extensiveness, an illusion of being-in-time such as is consistent with our sense of real people. The effect is similar to what J. J. Gibson, speaking of humans' intuitive inference of the solidness and continuance of physical objects, called "kinetic occlusion," whereby the movement of an object across a surface "progressively covers and uncovers the physical texture of [the object] behind it" (Scarry, *Dreaming by the Book* 12–13).

More interesting, though, than characters who talk occasionally about absent people, places, or events are those times in plays when the telling of stories is construed conspicuously as attempts at storytelling that seem deliberately to bypass conventional mimetic enactments. For whereas in the Nurse's speech in *Romeo and Juliet* the relationship between on- and offstage events is hidden by the "natural" revelations of character and by the knowledge that they took place before the "real time" of the representation, the speeches of the various gentlemen in *The Winter's Tale* point to a contemporary and intentionally occluded scenic enactment. Narrative in this case somehow draws attention to its iconic inadequacy with respect to what it purports to represent; we are asked to trade an actual sight for an imagined one. These kinds of ostensive uses of narrative brush against the grain of mimesis; rather than pass unnoticed, as was the case in the speeches by Lopakhin or Juliet's Nurse, they tend to confront us with evidence of the essential poverty of language as a substitute for sight. At such times, narrative, or "telling," seems to threaten mimetic cohesiveness, sometimes marginally or fleetingly, sometimes—as, for example, with the unusually long introductory monologue spoken by Egeon in *The Comedy of Errors*—to the extent that the narrative element becomes detached from its dramatic context and threatens to annul it.

Critical discussions about the difference between showing an action and telling about it begin with the dialogues of Plato, who distinguished two different kinds of poetic imitation: verbal diegesis, a mode of performance in which the Homeric bard recounted the action as it were from the third person, and theatrical mimesis, in which the poet switched from third to first person, modeling himself on the various characters and in effect becoming them in the manner of an actor. Plato's analysis seems to have been more theoretical than descriptive. It's unlikely that any dramatic performance that included choral elements could ever be fully mimetic in the way that Plato describes. Likewise it is hard to imagine that an epic poet ever recited a poem without occasional recourse to some kind of direct imitative enactment, pretending now and then to be Paris or Ulysses by adopting their words and miming their actions.

Yet Plato's idea that mimesis and diegesis were in some essential way contradictory proved as useful and popular as it was rigid and exclusive, and his account of these two differing modes of poetic performance seems to have been adopted as the basis for most subsequent comparisons between drama and narrative literary forms, especially—perhaps even surprisingly—among those commentators who attempt to rescue mimesis from Plato's attacks. Plato's distinction between diegesis and mimesis, for example, seems to have been accepted without question by Aristotle, one of his pupils, who in answering Plato in *The Poetics* assumes a similar essential difference between drama and epic poetry. In discussing the various possible modes of poetic imitation, Aristotle says that either "the poet may imitate by narration . . . or he may present all his characters as living and moving before us." And he subsequently underscores this fundamental opposition between epic poetry and drama, calling the latter an imitation "in the form of action, not of narrative" (Aristotle 20, 22). Likewise the Roman poet and critic Horace, writing in *The Art of Poetry* (c. 24 BCE), follows Plato and Aristotle in asserting that theatre shows action mimetically, whereas narrative represents it only through language. In distinguishing between showing an action, on the one hand, and telling about it, on the other, Horace takes Plato's argument against mimesis and turns it to drama's advantage: "Events are either acted out on the stage or else they are narrated. Now the mind is much less stirred by hearing things described than it is by actually seeing them, with one's own eyes, as a spectator" (Horace 54).

Not only does Horace suppose narrative and drama to be incongruous (if not in principle exclusive) but also, therefore, privileges "showing" for the very reasons that Plato scorned it, namely, its efficacy in causing spectators to credit the artistic illusion with truth. It is this view of the fundamental superiority of "showing" over "telling" that is handed down as part of the classical tradition of criticism of theatre. The status of drama as allied principally to the experience of seeing is further reinforced by the fact of its etymology. Lyric poetry takes its name from the musical instrument that was played to accompany its performance, while drama and theatre derive, respectively, from *dran* (a "doing") and from *theatron*, the "seeing place," namely, the place where the audience sat during dramatic competitions in ancient Athens. Even when drama is taken to be one of the several varieties of literary production (as opposed to a performance art), it has nevertheless been understood to be inherently visual. Indeed one contemporary theoretician of the genre calls drama "a visual art *par excellence* and an institutionalized space for voyeurism" (Pavis 388).

Normally proscribed from the stage, therefore, or considered at best a necessary concession to the various exigencies of theatrical performance, narrative is to be used in plays only sparingly and only when there is no mimetic alternative. This is the position taken by Aristotle, articulated more fully by Horace, and, by and large, it is the position adopted subsequently by the French neoclassicists, as, for example, when Corneille writes in his "Examen du *Cid*": "what is exposed to view is much more moving than what is learned only through a narrative." Indeed, since antiquity, most commentators on drama follow classical and neoclassical precepts in believing narrative to be at best a *substitute* for enactment, never its equivalent. As Patrice Pavis has recently put it, narrative cannot assume too great a role in drama "without running the risk of destroying its theatrical quality" (Pavis 230).

When Aristotle describes the relationship of direct enactment and narrative, he does not bother to discuss the many circumstances in Greek drama when telling takes the place of showing. But Horace does take up the question of the relationship of one to the other, and implicit in his commentary is a much subtler assessment of the two modes of performance. After stipulating the superiority of enactment over narration, Horace goes on to explain the differences between events that can be shown on stage and a class of actions that he says should instead simply be told about; while granting (as in the above citation from *The*

Art of Poetry) that showing is always more stimulating to the mind than telling, he cautions that

> [o]n the other hand, you must not show on the stage itself the kind of thing that should have taken place behind the scenes. In fact, many things must be kept from sight for an actor to tell about later. For example, Medea should not butcher her children in plain view of the audience, nor the wicked Atreus cook human flesh in public. Nor, of course, should Procne be transformed into a bird, nor Cadmus into a serpent. Whatever you try to show me openly in this way simply leaves me unbelieving and rather disgusted. (Horace 54)

In enumerating things that ought to be excluded from view, Horace cites several obvious monstrosities, and to the casual reader it sounds as if he wants dramatists to avoid depicting things such as butchery or cannibalism mainly on the grounds of good taste. In much the same spirit as Horace advises dramatists not to show Medea's savage cruelties, for example, contemporary television broadcasts typically walk a fine line between showing and telling. The cynics' view of television news is that "if it bleeds, it leads," but television reporting rarely shows images of bodies mutilated by war, accident, or criminal acts. In 1994, to take one well-known example, during the extensive coverage of the murder trial of O. J. Simpson, television viewers listened to meticulous descriptions of the victims' stab wounds but never actually saw police photographs of the corpses, even though these were readily available for broadcast. More recently, in connection with news about the mistreatment of inmates at Abu Gharaib prison in Iraq, one sometimes heard arguments that for the news media to show pictures would be irresponsible and potentially inflammatory, whereas detailed verbal descriptions were considered to be suitably informative.

Horace's examples can indeed be interpreted as a plea for decency or restraint in theatrical representation, and they have often been so construed. But there are other issues besides taste at play in *The Art of Poetry*, and here, as is true of many ancient critics, Horace is mindful of aesthetic as well as ethical criteria. A second and subtler level of Horace's argument seems to be concerned with credibility. Rather than attribute Horace's injunctions to decorousness, as that term is commonly understood, it may be more accurate to say that for Horace, "decorum" is essential for verisimilitude as well as for decency. In fact, if anything, his emphasis seems to be more on the former than the latter: "whatever you try to show me openly in this way *simply leaves me unbelieving* and rather disgusted" (my italics). In other words, Horace implies that cannibalism, theriomorphism, and child murder ought to be proscribed not only because they are abhorrent to watch but also because in some sense acts such as these are beyond mimesis; actors who attempt them in effect sabotage their own representing. Were he to write for a modern critical journal, Horace might put it this way: "Since nobody believes that an actor playing Atreus or Cadmus is really eating human flesh or becoming transformed into a snake, the audience is made fully aware of the awkward split between signifier and signified."

In writing this book I was chiefly interested to unpack the reasons narrative might be selected by a playwright in preference over scenic enactment, so that, for example, crucial elements of a play either remain unseen or, alternatively, are seen only as subjected to a narrative logic. The plays I discuss range from ancient to modern, although my main concern is with works of the twentieth century. My commentaries are for the most part not sustained interpretive readings of texts or parts of texts; I intend mainly to highlight some of the main ways in which playwrights have conceived of the relationship between narrative and scenic enactment and some of the different uses to which the space "offstage" has been put. I try to make the examples sufficiently numerous and varied to be illustrative, though they can hardly be exhaustive. It was not my intention to write a history; my aims were simpler and twofold. First, I wanted to try to draw attention to a part of drama and criticism of drama that has received relatively little attention; and next, I wanted also to show the widespread ascendancy of narration, or telling, as an alternative strategy to enactment or showing in twentieth-century drama. Now that the twentieth century is history—they aren't, as Will Rogers said in recommending the buying of land, making any more of it—one can more clearly see the broad aesthetic patterns that define a period.

Finally, let me say a word about the book's organization, which I chose to structure along thematic lines rather than by chronology or author. The book draws on plays and criticism from ancient as well as contemporary authors; it covers a broad spectrum of characters, events, or objects that are not represented visually, from the offstage murder of Agamemnon to the imaginary scenery of *Our Town*, from between-the-scenes sex in Arthur Schnitzler's *Reigen* to the elusive Mr. Godot. I start with the assumption that there is little if anything for which a competent dramatist cannot devise a suitable visual correlative, whether realistic or stylized, should he or she choose to do so. Representation in the visual arts is always a two-way street, as Ernst Gombrich long ago established, dependent not only on the skill of the artist but also on the willingness of the spectator to enter into a kind of contract such that a few strokes of light and color, or the barest sequence of words and gestures, function to bring a make-believe world to life. Enhanced by a willing beholder's imagination, a stick can become a horse, a pair of actors armed only with wooden swords can represent an entire army, or (as Bottom says in *A Midsummer Night's Dream*) with the application of a little plaster or loam, a man can easily stand in for a wall. If there are, in theory, few limits as to what *can* be represented on stage, then, it follows that the things that playwrights choose *not* to represent can be as significant as what meets the eye. My concern ultimately is to show how a deliberate turning away from scenic enactment sometimes forms a critical part of theatrical art.

The book is divided into three chapters; each of these chapters focuses on a distinctly different way that narrative elements have been used to stand in for or to offset direct mimetic enactment. Sometimes, as, for example, in the case of Egeon's expository speech to the Duke in *The Comedy of Errors*, narration substitutes for mimesis; in this case the playwright simply exchanges one mode of representation for another. But the relationship between narrative and enactment is not always this simple, and at other times telling takes on more complicated and substantive dramaturgical functions. Especially in the twentieth century, playwrights develop strategies for narrative according to which telling does not just replace showing but complicates or even subverts it.

In the first chapter, "Showing vs. Telling," I look at some of the ways in which narrative reports have been used by dramatists to replace mimetic enactments. I begin with a discussion of the role of messengers in ancient tragedy. These "messenger speeches," in which a character reports on things that are understood already to have taken place somewhere in the space beyond the mimesis, can be found everywhere in classical Greek theatre; indeed they are one of the cornerstones of tragic dramaturgy. But "messenger speeches" exist in abundance in modern works as well. These speeches, in which one or more characters come on stage to give a retrospective account of (usually violent) events that have not been enacted in view of the audience, are designed normally to make those unseen events present in the imaginations of spectators. For the most part, speeches such as these take place within an otherwise fully mimetic context; these narrators are almost always characters in their own right, and they generally have plausible motives and reasons to be where they are and to say the things they say. Other instances in which dramatists have historically substituted telling for showing involve the depiction of sex acts, and this chapter concludes with a comparison of several older, relatively discreet representations of sexuality with more recent styles of enactment.

In the second chapter, "Against Mimesis," I turn to plays in which the narrative elements are generally not subsumed within a broader mimetic strategy but are instead made to function more or less independent of the mimesis as components of a partly diegetic dramaturgy. These are typically modern or postmodern plays in which the narrative elements exist not for the sake of the mimesis but to carry out their own formal or thematic purposes. Here narrative is used not only to paint a picture of unseen events or to bring what is offstage to light, but also to qualify mimesis by drawing attention partly away from the represented object onto the codes of its representation. Fundamentally an "antitheatrical" strategy, the employment of narrative in these instances becomes a way to distance the theatre from itself, to subvert mimesis, as it were, or to disable or displace it.

In the third and last chapter, "Theatres of Absence," I examine a number of plays that seem to be built purposely around characters or events that are never permitted to be seen. Plays in which events can be said to be "represented" only by way of their absence, plays in which characters are nothing more than functions of words, might sound at first like avant-garde experiments. But plays like Samuel Beckett's *Waiting for Godot* are relatively recent instances of what seems to be an old theatrical strategy as well as a common modern practice. Twentieth-century playwrights as diverse as Susan Glaspell, Federico García Lorca, and Thomas Bernhard, all create dramas structured according to what might be called a poetics of omission. This final chapter takes the idea of a dramaturgy of the unseen to a logical extreme, where, paradoxically, the staged enactment, instead of being the embodiment of an absent reality, becomes instead the outward sign of (in Hamlet's words) "that which is beyond show."

CHAPTER 1
SHOWING VS. TELLING

It would be a mistake to think of the messenger's speech as a poor substitute which fails to make up for what cannot be shown on stage. On the contrary it is superior to spectacle.
—S. A. Barlow, *Euripides' "Trojan Women"*

Greek tragic playwrights typically dramatized violence or dying by means of so-called messenger scenes. Bustling on stage to report after the fact on events such as the slaughter of Clytemnaestra or the dismemberment of Pentheus, the messenger is one of the best-known conventions of ancient drama. One of the best-known and yet also one of the most hackneyed: the messengers' elaborate narratives of disaster figured so prominently in ancient tragedies that they soon became objects of ridicule, even among contemporaries. The comic poet Aristophanes often makes fun of the messengers of tragedy for their habit of making short stories long. In his earliest play (*Acharnians*, 425 BCE), for example, a messenger comes on stage to deliver a parody of a retrospective narrative concerning a battle; the climax of his report comes when he tells how the Athenian general Lamachus stumbled into a sewage ditch. And in *Lysistrata* (411 BCE), Aristophanes' best-known comedy, a play in which the women of Athens refuse to have sex with their husbands and lovers unless they agree to put an end to the Peloponnesian War, a messenger from Sparta appears on stage suffering from a stupendous penile erection, one so large and irrepressible it impedes his story-telling.

Modern spectators, especially those whose tastes and expectations with respect to the enactment of death or injury are in part shaped by a film industry that glories in making visible the wholesale destruction of biomass, sometimes find these restrained classical ways of representing violence to be tedious and insipid. Even a Victorian classicist such as A. E. Housman makes fun of the way the spectators of Greek tragedy have to rely on their ears rather than their eyes when events turn bloody. Housman's satiric "Fragments of a Greek Tragedy" mocks the Greek playwrights' periphrastic style as well as their habit of staging all violence out of sight, in this case (as in the *Agamemnon* of Aeschylus) behind the closed doors of the king's palace:

> *Eriphyla [within]*: O, I am smitten with a hatchet's jaw;
> And that in deed and not in word alone.
> *Chorus*: I thought I heard a sound within the house
> Unlike the voice of one that jumps for joy.
> *Eriphyla*: He splits my skull, not in a friendly way,
> Once more; he purposes to kill me dead.
> *Chorus*: I would not be reputed rash, but yet
> I doubt if all be gay within the house.
> *Eriphyla*: O! O! Another stroke! that makes the third.
> He stabs me to the heart against my wish.
> *Chorus*: If that be so, thy state of health is poor,
> But thine arithmetic is quite correct. (Olson 4–5)

Housman's parody is so funny because it is so close to the mark. Especially to beginning students of Greek tragedy, who are almost always at a loss as to how to respond appropriately to the typically inflated style of Aeschylus, the language really does sound that absurd. But naturally this cannot have been the customary reaction of audiences in the fifth century BCE. Vase paintings from the period often depict the very actions that the tragic playwrights chose not to stage, and frequently the scenes on the vases include marginal characters (typically, old men and slaves) who can be identified as the "messengers" belonging to the stage work. The presence of these figures within the overall pictorial composition suggests that the vase painters were depicting the events as they had been told about in the messengers' narratives—for example, Medea killing her children or Hippolytus being overwhelmed by a monster rising from the sea (Taplin 82). Since the events described in messengers' accounts appear fairly regularly on vases, it is fair to say that those scenes must have been present to the painters' eyes, even though they had never been actually staged, and so they must have been familiar to Athenians in general. Indeed those scenes may have been commonly "pictured" as part of Athenian myth, forming an important part of the fabric of popular culture, much like our own collective images of the Kennedy motorcade as it traverses Dealey Plaza in Dallas or marines raising the flag on Iwo Jima. The narrative component of the play, in other words, far from being secondary or aniconic, in some way must have been "seen" with particular vividness by numerous artists, and it would be wrong not to assume a similar response on the part of the majority of spectators.

Of course, there are important differences between ancient audiences and modern ones. Making and listening to oral narratives are not now—as once they were—an important part of social life. In preliterate societies, listening patiently to someone relate a story accorded fully with patterns and practices of daily life involving education, news gathering and dissemination, and politics. In ancient Athens especially, writes Sian Lewis, "business of all kinds, public and private, was conducted orally, rather than by written means....News could be spread within a *polis* either by public announcement, or by informal word of mouth" (Lewis 433). Thus when the city of Elateia fell in 339 BCE, Theophrastus, writing about the disaster, describes the scene in Athens when a herald arrived in the agora on the day after: "It was evening, and a messenger came to the prytaneis with the news that Elateia had been taken. Some of them straightaway got up from their dinner and drove the people from their stalls in the Agora, and set fire to the booths" (Lewis 434).

To the extent that theatrical conventions reflect the social practices of their specific cultures, then, one might expect that few if any plays from more literate societies would include characters whose only purpose is to relate the news. But in fact "messenger speeches" can be found often in plays from later eras, when cultural habits were (or are) no longer predominantly oral and when the classical influence is remote or nonexistent. Twentieth-century dramatists in particular make considerable use of retrospective narrative reports on the lines of this ancient mode of communicating news, belying its status as an outmoded convention. Even cinema, a medium much more centered on visual experience than drama, occasionally makes use of old-fashioned narratives to represent past events, even when it would be perfectly feasible to represent those events pictorially by means of flashbacks or other cinematic techniques. One well-known example would be Quint's "Indianapolis monologue" in the horror film *Jaws*. At sea, the first night out on the hunt for the great white shark that has been marauding the beaches on Cape Cod, the three main characters in the movie—Hooper, a marine biologist; Brody, the local chief of police; and Quint, a crusty fisherman and sometime shark hunter—have a drunken conversation about the sinking of the American heavy cruiser Indianapolis during World War II:

> *Hooper*: You were on the Indianapolis?
> *Brody*: What happened?
> *Quint*: Japanese submarine slammed two torpedoes into our side, chief. It was comin' back from the island of Tinian Delady, just delivered the bomb. The Hiroshima bomb. Eleven hundred men went into the water. Vessel went down in twelve minutes. Didn't see the first shark for about a half an hour. Tiger. Thirteen footer.... Sometimes that shark, he looks right into you. Right into your eyes. You know the thing about a shark, he's got... lifeless eyes, black eyes, like a doll's eye. When he comes at ya, doesn't seem to be livin'. Until he bites ya and those black eyes roll over white. And then, ah then you hear that terrible high pitch screamin' and the ocean turns red and spite of all the poundin' and the hollerin' they all come in and rip you to pieces.... On Thursday mornin' chief, I bumped into a friend of mine, Herbie Robinson from Cleveland. Baseball player, boson's mate. I thought he was asleep, reached over to wake him up. Bobbed up and down in the water, just like a kinda top. Up ended. Well... he'd been bitten in half below the waist. Noon the fifth day, Mr. Hooper, a Lockheed Ventura saw us, he swung in low and he saw us. He's a young pilot, a lot younger than Mr. Hooper, anyway he saw us and come in low. And three hours later a big fat PBY comes down and start to pick us up. You know that was the time I was most frightened? Waitin' for my turn. I'll never put on a lifejacket again. So, eleven hundred men went into the water, three hundred and sixteen men come out, the sharks took the rest, June the 29, 1945. Anyway, we delivered the bomb. (Ryono http://www.whysanity.net/monos/jaws.html)

Quint's story has little historical accuracy with respect to the wartime experiences of the men of the Indianapolis who spent four days adrift in the western Pacific in the summer of 1945. According to the testimony of two of the crew, most of the men who survived the torpedo attack but died in the days before being rescued died not from sharks but were gradually worn down from a lethal combination of exhaustion, exposure, and dehydration. Still other men became delirious, drank seawater to ease their desperate thirsts, and, ironically, died even more quickly of dehydration brought about by a toxic accumulation of salt. Naval documentation of the incident includes mention of fifty-six men (of a crew

of about 1,200) whose bodies on recovery appeared to have been mutilated by shark bites, but neither of the two survivors' accounts describes
attacks by sharks as a significant threat: "But honestly," said one person
whose testimony forms part of the official naval record, "in the entire
110 hours I was in the water I did not see a man attacked by a shark"
(Haynes http://www.eyewitnesstohistory.com/indianapolis.htm).

It is not surprising that Hollywood rewrote history in order to
frighten moviegoers with the dangers of swimming in the open ocean,
but it is worth considering why the filmmakers elected not to show
images of the carnage. Certainly nothing remotely comparable to this
scene (let alone Quint's speech) occurs in Peter Benchley's novel, on
which the film is based. The manner in which Quint's narrative assists
viewers in the process of mental image-making is interesting from an
aesthetic standpoint precisely because it is so obviously oriented toward
language rather than the kind of direct sensory experience one expects
from a film. Because at that point in the movie the terrible great white
shark has still not been shown on camera, it seems clear that Quint's
speech was invented with specific pictorial effects in mind, chiefly to
aid spectators in constructing what might be called the felt experience
of seeing a shark attack. By giving an "eyewitness account" of the fearful experience of hundreds of men adrift amidst a multitude of ravening
sharks, Quint's speech, like the speeches of the messengers of ancient
Greek tragedy, acts literally on the imagination, that part of the mind
that is used to construct virtual pictures of reality. We can understand
better the conditions within which such "messenger speeches" function
for modern audiences, therefore, if we review briefly some of the criticism on these lengthy narratives with respect to their suitability as an
alternative to dramatic enactment.

*   *   *

The history of criticism on messenger speeches in drama is not extensive, but what little there is has remained remarkably consistent
over time, from the classical period up to the present. At the center
of this ongoing commentary is the consciousness of a fundamental
tension between the two basic styles of representation available in the
theatre—enactment and narration. This tension between showing
and telling in turn derives from analogous differences in the artistic
modalities that underlie them, between, on the one hand, a mimesis
dependent mainly on visual imaging (as carried out by painters, for

example) and, on the other, the practice of ecphrasis (an extensive and exclusively verbal account of a visual image, in particular of a painting, but more generally, of any person or event). The argument concerns the relative adequacy of these two entirely different strategies of artistic representation. Horace is among the earliest writers on theatre to note how difficult it was for spectators at a play to believe in the mimetic adequacy of certain kinds of staged enactments, and his skepticism is again taken up by literary theorists of the Italian Renaissance. Lodovico Castlevetro (1505–1571), for example, extended Horace's suggestion that the appearance on stage of certain extreme kinds of violence tended paradoxically to generate their own exclusion, so to speak, from the mimesis. Castlevetro theorized further that any kind of violence—not just instances of unusual cruelty or brutality—ought not to be shown on stage because such acts could not be done with verisimilitude. Similarly Dryden (1631–1700) attempted to elaborate on why a few certain actions, in particular dying, ought to be told about rather than shown. "All *passions*," he says,

> may be lively represented on the stage…but there are many *actions* which can never be imitated to a just height: dying especially is a thing which none but a Roman gladiator could naturally perform on the stage, when he did not imitate or represent, but do it; and therefore it is better to omit the representation of it. (Dryden 143)

Dryden's argument, like that of Horace, proceeds mainly on the basis of verisimilitude and the limits of spectacle; he is particularly concerned to identify those kinds of actions or events that, if imitated on a stage, by their nature strain spectators' credibility beyond the breaking point:

> The words of a good writer, which describe it lively, will make a deeper impression of belief in us than all the actor can insinuate into us, when he seems to fall dead before us; as a poet in the description of a beautiful garden, or a meadow, will please our imagination more than the place itself can please our sight. When we see death represented, we are convinced it is but fiction; but when we hear it related, our eyes, the strongest witnesses, are wanting, which might have undeceived us; and we are all willing to favour the sleight, when the poet does not too grossly impose on us. (Dryden 143)

That is why Restoration audiences laughed at actors who pretended to die in front of them, and perhaps that is also why when *Schindler's*

List was first screened in 1993, Oakland teenagers laughed during some of the scenes involving mass executions. The laughter in this latter case (whether or not the youths were capable of articulating the reasons for laughing is beside the point) may have been predicated on classical dramaturgical principles rather than on contemporary, hostile race relations between Jews and blacks, as was often assumed at the time.

Horace and Dryden are fully aware of such subtle problems of theatrical representation, and many of their arguments concerning verisimilitude can be found essentially unchanged in newer and seemingly more sophisticated theories of drama. Compare, for example, Dryden's analysis of the inefficacy of death scenes on the Restoration stage with Anthony Kubiak's *Stages of Terror* ,a recent and highly abstract discussion of the depiction of violence in contemporary theatrical performance. In writing about the late twentieth-century phenomenon of performance artists such as Chris Burden, who arranged to have himself shot (*Shoot*, 1971), or Vito Acconci, who sat naked in front of a camera and bit himself on those parts of his body he was able to reach with his teeth (*Trademarks*, 1970), Kubiak suggests that violence can never be presented as theatre without at least partly betraying it. The attempt to introduce real violence into theatrical performance, according to Kubiak, is not "artistically efficient." "There have been artists," he writes, "Valie Export and Stuart Brisley, for instance—who have investigated this very impasse in their work. Yet even these inquiries have fallen short, because when actual violence is mediated in performance, when actual violence becomes 'present,' it inevitably shifts the emphasis of the problematic onto the *inaccessibility* of real, but mediated, pain" (Kubiak 144).

Kubiak's argument rests on a premise similar to that of Dryden and Horace: spectators' prior knowledge of real pain, one might say, shrinks the range of actions that can be performed by an actor whose apparent suffering and pain—even if they are, in fact, real—will be understood within the theatrical frame as only the structure of those realities, visibly enacted. What was originally an essential component of the world and of human experience (pain, suffering, dying, or almost any kind of violence), in other words, now seems translated into theatre. Kubiak writes that "here we are faced with a paradox, however, because in performance what cannot be articulated must be shown, and when it is shown, it ceases to be what it was. Thus when terror enters the information systems of performance, it ceases, in a sense, to be terror—which is unspeakable, and unrepresentable—and becomes a mask of itself" (Kubiak 11). "Pain and terror," he concludes, "become seemingly less real—even, in their repugnance, more easily dismissable—as they are anchored in the Symbolic and become purely codified and commodified" (Kubiak 145).

In the back of all these various assertions about what can or cannot be shown effectively on stage, therefore, lies a novel and important consequence for theatrical representation: in certain cases, narrative may be better suited than enactment to cause an audience to picture events. The issue then turns on the difference between a representation based on direct scenic enactment and one that instead relies on what Scarry calls "dreaming by the book," a process of verbal-based imagining whereby the writer aids the reader (in the case of messengers' narratives, the viewer) in the act of mental picturing by means of a series of written or spoken cues and instructions.

Spectators' imaginings, of course, are always a part of theatre, in the most trivial as well as in the most profound sense; even amateur actors have no trouble in using words to help audiences see swords, moonlight, and landscapes, things that are not really materially represented before them. The imagination, says Frank Kermode, is "a form-giving power, an esemplastic power"; whether its products appear true to existence or in shapes "preposterously false...they change with us, and every act of reading or writing a novel is a tacit acceptance of them" (Kermode 144). In every act of seeing or writing a play, one might add: the potential *enargeia*—the power of language to create a vivid presence of that which is set forth in words—that can be tapped by a skilled dramatist is easily demonstrated with regard to the mise-en-scène, where substituting verbal accounts for cloth and paint does not restrict spectators' sense of location and may actually enhance it. When W. B. Yeats remarked of his own progress as a dramaturg that "it was a great gain to get rid of scenery," he was celebrating the triumph of the imagination as it attends and completes the actors' mimesis. Likewise, Thornton Wilder insisted that the narrated scenography of *Our Town* was superior to any possible actualization, and the stage history of Wilder's masterpiece proves him to have been right. Early productions in Princeton and Boston with conventional naturalistic décor were unsuccessful. Audiences began to respond enthusiastically to *Our Town* only after the show moved to New York, when Wilder persuaded his producer to throw out everything in the set except for some chairs, a table, and a couple of ladders. Since then the published text has begun with a bold injunction: "No Curtain. No Scenery."

The successful performances of these and other plays similarly dependent on spectators' imaginations are perfectly consistent with contemporary research findings on the mechanisms of visual perception and the visual imagination. "Imagination," says the psychologist Daniel Gilbert, "is usually effortless. . . . Like perceptions and memories, these mental pictures pop into our consciousness *fait accompli*. We should be grateful for the ease with which our imaginations provide this useful service, but because we do not consciously supervise the construction of these mental images, we tend to treat them as we treat memories and perceptions—initially assuming they are *accurate representations* of the objects we are imagining" (Gilbert 98, 99).

But what is it that dramatists actually do to create in the minds of the spectators a set of verbal instructions such that the images and actions they call to mind are in some respects the equivalent of direct sensory perception (if not, as some testimony implies, actually its superior)? Messengers' graphic narratives belong to the more general literary topos *ecphrasis*, a rhetorical exercise that in its restricted original sense referred to a verbal description of a painting, but which in broader application could be (and was) applied to any use of descriptive language such as could bring about an imaginative visual experience that in effect made hearers into spectators (Zeitlin 157). Something like ecphrasis is the effect of the lyrics sung by the Chorus as they come on stage in Euripides' *Ion*, for example; their words consist of a highly detailed account of some of the decorations on the temple of Apollo at Delphi, taking the form of specific instructions to one another (and, by extension, to spectators also) to picture images that in all likelihood were not actually present on stage during the performance as part of Euripides' mise-en-scène:

> Look, look at this: Zeus's son
> Is killing the Lernaean Hydra
> With a golden sickle.
> Look there, my dear. . . .
> And look at this one
> On a horse with wings.
> He is killing the mighty three-bodied
> Fire-breathing monster.
> My eyes dart everywhere.
> See! The battle of the giants

On the marble walls.
Yes, we are looking. (Euripides, *Ion* 191)

This kind of elaborate depiction of physical details of a particular locale is common on the classical façade stage (and in presentational forms of theatre in general). Such descriptions, as Froma Zeitlin writes, serve multiple purposes: they are verbal exercises designed not only to create the theatrical illusion of an absent presence in the minds of the spectators but also to bring the felt immediacy of direct scenic enactment, which is the normal mode of theatrical representation, into high contrast with a different epic or diegetic technique, what Zeitlin calls "the symbolic potential of a figured design" (Zeitlin 149).

To see further how ecphrastic narrative forms the dramaturgical basis of an "aesthetics of abstinence" (Steiner 227), consider an exemplary messenger speech from Greek tragedy—the description of the deaths of Creon and his daughter in Euripides' *Medea*. According to legend, Medea was a barbarian princess and sorceress, the daughter of King Aeetes of Colchis who kept the Golden Fleece. She fell in love with Jason, the leader of the Argonauts, and helped him to steal the Fleece and subsequently to escape. The pair fled first to Iolchus, Jason's hereditary kingdom, subsequently to Corinth, where Jason, in order to advance his political fortunes, abandoned Medea and announced his intention to marry the daughter of the king of Corinth. When the play begins, Medea, jealous and enraged, decides to take her revenge by murdering Jason's two children. She is motivated both by the desire to wound Jason and by the perverse wish to spare her children the hardships they would endure when she could no longer be certain of their safety. Medea also murders the princess as well as the princess's father, Creon. It is a ghastly crime: the bodies of both people are consumed by a slow-burning poison that Medea had concealed in a crown and robe she gave to the princess as wedding presents. These agonizing deaths are not enacted on stage but (as is customary in Greek tragedy) narrated in vivid detail by a messenger who claims to have seen what happened. The messenger's speech continues uninterrupted for nearly a hundred lines of verse, but Euripides supplies him with only a minimum of motive for telling so long a story. When he arrives on stage, Medea simply asks him to tell her how her victims died, sadistically requesting him "not to hurry his account." Even though Medea's "need to know" provides a plausible motivational frame for the messenger's speech, it is nevertheless clear (as it must have been to Euripides' original audience) that the real reason for this particular scene was dramaturgical rather than psychological. Here is an excerpt from the messenger's narrative:

> [Our mistress], when she saw the dress, could not restrain herself.
> She agreed with all her husband said, and before
> He and the children had gone far from the palace
> She took the gorgeous robe and dressed herself in it,
> And put the golden crown around her curly locks,
> And arranged [*schematizetai*, "arranges"] the set of the hair in a
> shining mirror,
> And smiled at the lifeless image of herself in it.
> Then she rose from her chair and walked [*dierchetai*, "parades
> through"] about the room,
> With her gleaming feet stepping most soft and delicate,
> All overjoyed with the present. Often and often
> She would stretch her foot out straight and look along it.
> But after that it was a fearful thing to see.
> The color of her face changed, and she staggered back,
> She ran [*chorei*, "flees"], and her legs trembled [*psthanei*, "trembles"]
> and she only just
> Managed to reach a chair without falling flat down.
> An aged woman servant who, I take it, thought
> This was some seizure of Pan or another god,
> Cried out [*ora*, "cries out"] "God bless us," but that was before she saw
> The white foam breaking through her lips and her rolling
> The pupils of her eyes and her face all bloodless.
> Then she raised a different cry from that "God bless us,"
> A huge shriek, and the women ran, one to the king,
> One to the newly wedded husband to tell him
> What had happened to his bride; and with frequent sound
> The whole of the palace rang as they went running.
> One walking quickly round the course of a race-track
> Would now have turned the bend and be close to the goal,
> When she, poor girl, opened her shut and speechless eye,
> And with a terrible groan she came to herself.
> For a twofold pain was moving up against her.
> The wreath of gold that was resting around her head
> Let forth a fearful stream of all-devouring fire,
> And the finely woven dress your children gave to her,
> Was fastening on the unhappy girl's fine flesh.
> She leapt up [*pseugei*, "leaps up"] from the chair, and all on fire
> she ran,
> Shaking her hair now this way and now that, trying

To hurl the diadem away; but fixedly
The gold preserved its grip, and, when she shook her hair,
Then more and twice as fiercely the fire blazed out.
Till, beaten by her fate, she fell [*pitnei*, "falls"] down to the ground,
Hard to be recognized except by a parent.
Neither the setting of her eyes was plain to see,
Nor the shapeliness of her face. From the top of
Her head there oozed out blood and fire mixed together.
Like the drops on pine-bark, so the flesh from her bones
Dropped away, torn by the hidden fang of the poison.
It was a fearful sight; and terror held us all
From touching the corpse. [Euripides, *Medea* 98–100]

A great deal of the messenger's account consists of an elaborate description of events; it is a shocking, even horrific speech, remarkable for its attention to vivid and stomach-turning details. It seems self-evident that Euripides' purpose in writing such a long narrative is twofold, to inform the audience of certain events that they have not witnessed first hand, and to provide them with a sequence of verbal instructions for seeing the destruction of the princess and her father with their imaginations. Thus within the messenger's narrative there are numerous passages that mingle action and objective visual language, such as, for example, when he describes how the wreath of gold that was resting on her head "let forth a fearful stream of all-devouring fire," or, "from the top of her head there oozed out blood and fire mixed together." Naturally such language can be expected to have its perceptual counterparts in spectators' imaginations, even assuming one has not seen wreaths of gold or a head oozing blood and fire, and it is easy to make mental pictures consistent with the messenger's words.

But these more or less direct descriptive formulations are augmented with sequences of carefully constructed images or tropes, and these latter tend not just to reference the fictional world of Jason and Medea but also to be drawn instead from what must have been the common experiential world of Euripides' audience. A reference to "the white foam breaking through her lips and her rolling the pupils of her eyes," for example, would likely be consistent with experiences or sights with which Euripides' audience would have been familiar. The symptom of foaming at the mouth is mentioned by Hippocrates in his writing on epilepsy, for example; he notes that sufferers of "the sacred disease" sometimes foam and sputter at the mouth like dying persons. Or perhaps Euripides may have been calling to mind the frothy spittle that appears on the muzzle of desperately fatigued animals. Likewise an audience with firsthand knowledge of the processes of butchery or open hearth cookery would surely have had little trouble summoning up concrete memories of how roasted flesh drops away from bones. In either case the playwright turns spectators' attentions momentarily to their memories of real events in order that they might respond more fully to imaginary ones.

It makes sense in such a case to think of the messenger's narrative as a blueprint or set of instructions for how to imagine actual sensory content; the messenger's words combine the detail and vividness necessary for the mind to take a fake look at something that isn't really there. As a result of Euripides' unflinching account of the sheer gross physicality of death by poisoning, over the years many different generations of audiences seem to have been persuaded not only that they are seeing these absent events but also that they have engaged them with the strongest possible affect. This experience is consistent with recent scientific psychological studies in the neurophysiology of the act of imagining; the psychologist Daniel Gilbert explains the process:

> If I were to ask you whether a penguin's flippers are longer or shorter than its feet, you would probably have the sense of conjuring up a mental image from airy nothing and then "looking" at it to determine the answer. You would feel as though a picture of a penguin just popped into your head because you wanted it to, and you would have the sense of staring at the flippers for a moment, looking down and checking out the feet, glancing back up at the flippers, and then giving me an answer. What you were doing would feel a lot like seeing because, in fact, it is. The region of your brain that is normally activated when you see objects with your eyes—a sensory area called the visual cortex—is also activated when you inspect mental images with your mind's eye. The same is true of other senses. For instance, if I were to ask on which syllable the high note in "Happy Birthday" is sung, you would probably play the melody in your imagination and then "listen" to it to determine where the pitch rises and falls. Again, this sense of "listening with your mind's ear" is not just a figure of speech (especially since no one actually says this). When people imagine sounds, they show activation in a sensory area of the brain called the auditory cortex, which is normally activated only when we hear real sounds with our ears. (Gilbert 129–30)

But what can it mean to say that such pictures in the mind's eye are seen with equal or even, as some have argued, greater vivacity than concrete objects? What is not seen is by definition not seen, and it seems—on the surface at least—somewhat misleading to argue that a linguistic presentation should be "seen" more clearly than a visual one. In many ways, language is ill-equipped to substitute for sight. Its inadequacy in this respect is readily demonstrated; almost any attempt to describe a natural object in all its complexity will prove how ungracefully words stand in for vision. Take, for example, this extensive description of the texture, shape, and coloration of a flower blossom: "ruffled, slightly creped deep burgundy with slightly raised rib of same color, blue violet veining and shading, slight magenta eye, gold throat and slightly green heart. Beautiful satin finish. Sepals recurve giving blooms flat triangular look" (Scarry, *Dreaming* 58). This is by any standard a remarkably detailed instance of ecphrasis, but few people who read it (it is taken from a flower catalog) would be likely to think that they had been given a satisfying substitute for even the briefest glimpse of the daylily known as "Satinique."

How to assess the relative efficacy of seeing Medea's handiwork with the mind's eye is a question, therefore, I want to consider at some length. In the first place, it is necessary to remember that the physical act of seeing is not a simple biological event, like respiration or digestion. It is an astonishingly complex amalgam of neurophysiological processes and cultural habituation, and even the simplest act of visual perception requires using what the physicist Arthur Zajonc calls an "inner light" (Zajonc 5). The role the mind plays in the act of seeing has been a matter of great controversy, ever since the Irish philosopher William Molyneaux in 1688 wrote to John Locke asking what would be seen by a blind man suddenly gifted with the power of sight. Would such a person, Molyneaux wondered, having gained prior tactile experience of a globe or a cube, be able to recognize the objects on seeing them for the first time? On the basis of recent evidence, the answer seems no: interviews conducted with patients who were blind from birth by cataracts but whose sight had been restored by surgery when they were adults seem to suggest that innocent eyes do not see the world with anything like normal pictorial competence. Concepts such as form, distance, and size in particular seem to be meaningless: "It would be an error to suppose that a patient whose sight has been restored to him by surgical intervention can thereafter see the external world. They eyes have certainly obtained the power to see, but the employment of this power, which as a whole constitutes the act of seeing, still has to be acquired from the very beginning. . . . To give back sight to a congenitally blind person is more the work of an educator than of a surgeon" (von Senden 160).

To say that messengers' narrated accounts of offstage events substitute for direct scenic enactment, therefore, is at best an oversimplification, at worst a misunderstanding of the nature of the transaction dramatists make when showing is replaced by telling. The function of the messenger speeches is not so much to replace sight by a literary sleight of hand as to stimulate a mental seeing of a somewhat different and more complex order. What takes place is not a simple point for point exchange of visual data for their equivalents in language, but a kind of psychogenesis whereby only a few key stimuli are used by the mind to construct what then seem like real memories and fully complete visual perceptions. This habit of treating sketchy verbal descriptions as if they were accurate sensory experiences of things themselves may be part of the hardwiring of human brains when it comes to processing symbolic representations; "ideas present," as an old Chinese formulation puts it, "brush may be spared performance" (Gombrich 331). The frequency with which ancient playwrights relied on messenger narratives may also have been related to the development in contemporary Athens of what Simon Goldhill has called a unique "culture of viewing," a culture where the relationship between visual and verbal signs was particularly fluid and where, in Simonides' famous dictum, "painting is silent poetry and poetry is painting that speaks" (Goldhill 197; Zeitlin 162).

Another consideration when attempting to assess the pictorial qualities of Euripides' messenger speech (or almost any other messenger speech, for that matter) has to do with its distinctive editorial quality. The speech is given as a narrative, a linguistic medium that filters experience according to various signifying conventions or interpretive codes and that, according to narratologists, is "by definition a discourse informed by a retrospective intelligibility" (Fleischman 32). The messenger's description, therefore, modifies and complicates the technique of ecphrasis. He reports not as a spectator to something he is witnessing (as in the case of the Chorus in *Ion* who report on the architectural decorations before which they stand), but to something he has seen previously and is recalling from memory. This backward-looking orientation of his narrative surely influences its reception by an audience. Because he relates events that are understood to have already happened somewhere "offstage," Euripides' messenger functions as a kind of editor or historian. In telling his story he makes the actions and people who are the subject of his narrative susceptible to distancing and an analysis, presenting them to the imagination in a relational rather than an unmediated way. "Narration," says Suzanne Fleischman, "is a verbal icon of experience viewed from a *retrospective* vantage; the experience is by definition 'past'" (Fleischman 23).

One could say that in the process of experiencing for himself the terrible effects of Medea's sorcery, the messenger began to perceive the possible story form that those events might take, and that the story as he relates it, then, is substantially different from the events as they might actually have occurred. It is clear, for example, that like any historian, Euripides' messenger uses the medium of narrative to govern spectators' responses in a way not possible with direct scenic representation. In the first place, he presents himself not only as an eyewitness to events but also as a commentator. He functions, in other words, less like a kind of surveillance camera that can be played back for viewers after the events in question have taken place, and more as an interpreter in his own right. On numerous occasions the narrator freely and openly uses his mind as a kind of critical screen, and these comments do not directly facilitate the mental creation of a simple picture. He refers, for example, to "an aged woman who, I take it, thought this was some seizure of Pan or another God, who cried out 'God bless us'"; he states bluntly that the corpse of the dead woman was "hard to be recognized except by a parent" and he concludes with his bleak opinion that "our human life I think and have thought a shadow." By introducing his own subjective interpretive responses into the narrative, by freely and openly using his mind, the messenger acts as a kind of "emotional filter" or "exegetical medium" (de Jong 77, 164), channeling the responses of spectators in a way not possible with direct enactment.

At other times the messenger frames the action in ways that bear even less resemblance to ecphrasis. Even though in many passages the messenger's language is expressly visual, his description does not seem necessarily consistent with what could be verified, say, by a photograph or by literal documentation. Instead the servant introduces highly subjective images into his account, and it is in part such idiosyncratic elements of his description that help spectators picture it as something once seen and now remembered. The description of the woman's head on which blood and fire merge "like drops on pine-bark," for example, does not so much substitute for sight of the mutilated body as it stimulates "seeing it" by way of the comparison with the scarred surface of a tree. Similarly, later in the narrative, the distraught father sticks to his daughter's dress "as the ivy clings to the twigs of the laurel." The body of the imagined man pictured as a consequence of that sentence does not stop at the usual limits of the human corporal form but instead extends out into the image of a pliant vine entwining itself over and around the limbs of the plant to which it clings for life and for support. Through such interpretive commentary the messenger continually modifies a listener's empathic involvement with characters and events they cannot actually see. Being asked to call to the mind's eye the sight of beads of pitch oozing from the bark of a tree, or strands of ivy twisted round the branches of the laurel bush, the imagination piggybacks on the neural machinery of actual perception. Spectators, accordingly, will throw their attention onto a quality of their own particular mental picture they had not hitherto noticed, and they will then "see" the dying woman in ways not possible were the scene to be portrayed by an actor present there before their eyes. Like a transceiver newly tuned to broadcast a subtler range of frequencies, the messenger's words stimulate an audience to feel a range of sorrow and helplessness about a situation they had previously not known they could feel so deeply about. The result is a theatrical experience that privileges mind over matter; what is more, this deliberate removal of violent physicality from sight, as George Steiner argues, can even give to the offstage world "a paradoxically intense nearness and pressure" (Steiner 227–28).

One additional rhetorical feature of Euripides' messenger's account suggests another way in which words, artfully arranged, call to mind a vivid image that can be as powerfully absorbing as direct enactment. A distinguishing stylistic feature of the messenger's language is that it contains a mixture of tenses, so that he renders his narrative partly in the past tense and partly using present-tense verbs. The speech as a whole contains ten instances of what is commonly called the "historic present" tense (de Jong 39), and these ten verbs, embedded within the larger context of the messenger's story involving a recitation of past events, complicate the overall pattern and sequencing of his narrative. Thus in describing the scene in the household just before the poisoning, the messenger begins his tale in the conventional narrative past (preterite) but soon switches briefly but emphatically into the present tense and then back again:

> When those two children, born of you, had entered in,
> Their father with them, and passed into the bride's house,
> We were pleased, we slaves who were distressed by your wrongs.
> All through the house we were talking of but one thing,
> How you and your husband had made up your quarrel.

Some kissed [*kusein*, "kiss"] the children's hands, and some
  their yellow hair,
And I myself was so full of my joy that I
Followed the children into the women's quarters. (Euripides, *Medea* 98)

Throughout his narrative the messenger alternates in this fashion between past and present tenses. Thus in the passage previously cited, once the poisoned crown has been placed on the princess's head, the narrator states that she "arranges" her hair. Then, having looked at her image in a glass, she "rises" from her chair and "parades through" the room. And later, when the messenger reaches that portion in his story when the poison has begun to take full effect, he again renders past events using the present tense: the princess "flees", "trembles," and ultimately "falls" to the ground.

The historic present in its simplest form represents bygone events from the point of view of someone who reports those events as if they were occurring simultaneously with the telling of them, much as in a radio broadcaster's account of a baseball game. A majority of people who have studied the use of the present tense as a narrative mode agree that its effect, generally speaking, is to create the impression of vividness or "breathless immediacy" (as *The Columbia Guide to Standard English* puts it), as if the events were taking place "before one's eyes" even as they are being recounted. Yet there is a subtle but important distinction between a sportscaster's reportage and a narrator's occasional, rather than consistent, use of the historic present. Obviously in attempting to render the past as if it were somehow present, the teller, unlike the broadcaster who describes a live event, must selectively alter his or her relationship to the story material; the narrator's tale begins in the past ("once upon a time…") and remains so framed, even though in switching tenses certain past events can be made to seem contemporaneous within an overall historical context. For this reason, contemporary narratologists sometimes prefer the technical term "presentification" (a translation of the German grammatical term *Vergegenwärtigung*) to describe the Euripidean messenger's style of mixing past and present tenses as an overall narrative strategy.

Common to most languages, the employment of "presentification" by speakers may have its origins in spontaneous oral narratives (Fleischman 55). The technique seems designed not simply to render past actions as if they were actually taking place in the speaker's (and listeners') present, but rather to "presentify" some acts or events that are otherwise acknowledged as existing in a historical (i.e., past) context. Some linguists distinguish further between a sustained use of the historic present as a consistent narrative strategy, on the one hand, and, on the other, the occasional and spontaneous use of the present tense within an otherwise conventional narrative, as is the case with Euripides' messenger's speech. In other words, what is important in this latter case is not the verb tense alone, but also how the shifts back and forth between past and present tenses determine the narrator's changing stance toward his or her material. "Tense alteration," says Monika Fludernik, "never signifies *temporally* as such, but signifies in relation to the narrator's (enunciator's) present" (Fludernik 262).

Euripides in this way exploits the historic present tense not only for its visual (or mimetic) potency but also for an additional diegetic purpose, namely, the ironic and retrospective highlighting of apparently trivial acts. All the things that the princess does are performed within a metaphysically expansive domain that lets hearers understand one dismal fact: that in everything she does, even as she is pictured in the act of performing a harmless gesture such as "stepping most soft and delicate," she is bound to die. If narrative, as Fleischman suggests, "is itself the 'cognitive instrument'…for imposing a structural coherence on the otherwise inchoate substance of experience" (Fleischman 95), then the servant's occasional use of present-tense verb forms does not simply cause an audience to picture the princess in the acts of rising, walking, or glancing in a mirror. Rather, by locating those imagined "present" acts in the context of events that have already taken place, the narrative implicates the princess in tragic circumstances where her every move, even the most seemingly inconsequential, only underscores her imminent fate.

The jolt of recognition caused by the messenger's selective "presentification" is breathtaking, as the retrospective narrative framework causes the unseen characters' present-tense acts to reverberate with their own tragic futurity. A purely mimetic reconstruction of the princess's suffering might well move spectators to pity and terror. But it would not be accompanied by spectators' sense of their own powerlessness to act to change what they know has already come to pass. As if with some cruel distortion of the technology of instant replay, spectators understand that the events they "see" occurring in their imaginations are, in fact, already over.

* * *

The practice of narrating scenes of violence in the classical manner is fairly common throughout the twentieth century. Even naturalistic playwrights sometimes rely on narrators to dramatize scenes of death and violence. *The Weavers* (1892), Gerhart Hauptmann's late nineteenth-century work, for example, though replete with numerous crowd scenes and with historically accurate sets depicting the living quarters of Silesian workers during the bloody uprising of 1844, defers almost all of its actual bloodshed to the offstage. Exactly as in Greek tragedy, one hears of offstage commotion indicating violence in Hauptmann's play—shots are fired, people scream, and so on—but none of this is enacted on stage. Instead Hauptmann structures his scenes according to the principles of classical dramaturgy: various characters run off and on stage, reporting on the murder and mayhem that has happened out of sight of the audience. Also in the classical manner, the bodies of the dead and the dying are subsequently brought on stage for public viewing:

> *A Woman Weaver*: (*Calls in*) Gottlieb, look at your wife. She's got more courage than you. She's jumpin' around in front of the bayonets like she was dancin' to music. (*Four men carry a wounded man through the entrance hall. Silence. A voice is clearly heard saying, "It's Weaver Ullbrich." After a few seconds, the voice says again, "He's done for, I guess—a bullet got him in the ear." The men are heard walking up the wooden stairs. Sudden shouts from outside, "Hurray, hurray!"*) (Hauptmann 156)

It is true that in the final scene of the play one character, Old Hilse, dies on stage. But the old weaver's death is not the focal point of the scene; Hauptmann depicts his death almost incidentally, rather like collateral damage. Seated indoors at work, Hilse is fatally shot by a stray bullet and dies without speaking, slumped against his loom. It is his daughter's and granddaughter's realization that the old man is dead rather than resting that constitutes the emotional meaning of the final moments of the drama; the effect is visually similar to the depiction of death by some of the realist painters of the late nineteenth century, where the dead or dying are pictured anonymously, almost banally:

> *Mielchen*: Grandpa, Grandpa, they're drivin' the soldiers out of town. They've attacked Dittrich's house. They did like at Dreissiger's. Grandpa! (*Frightened, the child sees that something is wrong—sticks her finger in her mouth and cautiously steps close to the dead man.*) Grandpa!

Mother Hilse: Come now, Father...say something! You're scarin' me!
(Hauptmann 156)

Sean O'Casey is another playwright of the early modernist period who often employs variations of the classical messenger speech. In his first play, *The Shadow of a Gunman* (1923), the key event is a nighttime raid on a Dublin tenement by members of the Royal Irish Constabulary. Minnie Powell, one of the residents of the tenement, hides her neighbor's homemade bombs in her own room in hopes that the Black and Tans would not think to search for them there. But they do. Minnie is arrested, but as she and the auxiliaries leave the tenement they are ambushed and Minnie is killed accidentally. None of these events is enacted, though they constitute the climax of the drama. Instead they are narrated almost as soon as they occur by a Mrs. Grigson, another of the tenement's residents, who rushes off stage to find out what's happening and then returns to give her account. After shots and explosions are heard offstage, Mrs. Grigson comes back on stage, excited and semi-hysterical, and, using many of the same narrative techniques for the representation of violent events that were used by the messengers of the classical stage, tells what happened to Minnie:

> Oh, Mr. Davoren, isn't it terrible, isn't it terrible! Minnie Powell, poor little Minnie Powell's been shot dead! They were raidin' a house a few doors down, an' had just got up in their lorries to go away, when they was ambushed. You never heard such shootin'! An' in the thick of it, poor Minnie went to jump off the lorry she was on, an' she was shot through the buzzom. Oh, it was horrible to see the blood pourin' out, an' Minnie moanin'. They found some paper in her breast, with 'Minnie' written on it, an' some other name they couldn't make out with the blood; the officer kep' it. The ambulance is bringin' her to the hospital, but what good's that when she's dead! Poor little Minnie, poor little Minnie Powell, to think of you full of a life a few minutes ago, an' now she's dead! (O'Casey, *The Shadow of a Gunman* 129–30)

The way in which O'Casey uses narrative to enrich understanding of an offstage event can be better understood by comparing the foregoing scene with a similar scene involving death by gunshot from another of his plays, *The Plough and the Stars*. This slightly later work (it was first performed in 1926) documents events from the lives of the residents of a Dublin tenement in the months leading up to the Easter uprising of 1916. The play includes a number of instances in which gunfire or artillery explosions are heard offstage, and these moments of imagined violence are typically followed by scenes in which characters come on stage to give retrospective accounts of scenes of death and destruction. In most of these situations, as was often the case in *The Shadow of a Gunman*, O'Casey again relies on a classical dramaturgy of retrospective narrative for the depiction of violence. But *The Plough and the Stars* also includes a final, climactic scene in which a fruit vendor named Bessie Burgess is shot while she is standing at a window. She then dies in full view of the audience. Here are O'Casey's stage notes and dialogue for part of the scene:

> [*With a great effort* Bessie *pushes* Nora *away from the window, the force used causing her to stagger against it herself. Two rifle shots ring out in quick succession. Bessie jerks her body convulsively; stands stiffly for a moment, a look of agonized astonishment on her face, then she staggers forward, leaning heavily on the table with her hands.*]
>   Bessie (*with an arrested scream of fear and pain*). Merciful God, I'm shot, I'm shot, I'm shot!...Th' life's pourin out o' me! (*To* Nora) I've got this through...through you...through you, you bitch, you!...O God, have mercy on me!...(*To* Nora) You wouldn't stop quiet, no, you wouldn't, you wouldn't, blast you! Look at what I'm afther gettin', look at what I'm afther gettin'...I'm bleedin' to death, an' no one's here to stop th' flowin' blood! (*Calling*) Mrs. Gogan, Mrs. Gogan! Fluther, Fluther, for God's sake, somebody, a doctor, a doctor! (O'Casey, *The Plough and the Stars* 215–16)

The foregoing passage represents less than half the entire episode as it is dramatized by O'Casey. To perform the scene in full requires at least several more minutes. After Bessie is shot and makes the previous speech, the distraught Nora calls for help in a breathless whisper to her recently deceased husband. Then Bessie staggers part way across the stage to the door, where she has the opportunity to make three more substantial speeches. First she laments her bad luck and misjudgment:

> This is what's afther comin' on me for nursin' you day an' night....I was a fool, a fool, a fool! Get me a dhrink o' wather, you jade, will you? There's a fire burnin' in me blood! (*Pleadingly*) Nora, Nora, dear, for God's sake, run out an' get Mrs. Gogan, or Fluther, or somebody to bring a doctor, quick, quick, quick! (*As* Nora *does not stir*) Blast you, stir yourself, before I'm gone! (O'Casey, *The Plough and the Stars* 216)

Up until this point the scene has been hectic, even chaotic. But as life ebbs from her, Bessie's speech becomes quieter, less abusive. In attempting to depict her death as tragic, it is no accident that O'Casey has Bessie use a variation of one of the most common tropes in classical tragedy for the experience of dying—the association of death with the vanishing of light and sightedness: "Jesus Christ, me sight's goin'! It's all dark, dark! Nora, hold me hand!" (O'Casey 216). Almost immediately Bessie sinks to the floor, and, as she takes her last breaths, she sings feebly the refrain from the well-known revivalist hymn "There Is a Fountain Filled with Blood":

> I do believe, I will believe
> That Jesus died for me;
> That on th' cross He shed His blood
> From sin to set me free.... (O'Casey 216)

Because this scene is in some respects a fully mimetic counterpart to the frantic messenger speech of Mrs. Grigson in *The Shadow of a Gunman*, it is worth considering what is gained or lost by having Bessie die over such an extended period of time and always in view of the audience. Certainly as O'Casey structures it, this climactic scene in *The Plough and the Stars* makes full use of the histrionic talents of the actor who takes the part of Bessie Burgess, calling into play both verbal and nonverbal modes of expression. Just to perform the scene adequately demands the actor convincingly express a variety of affects, including astonishment, fear, pain, anger, and regret—and that sweeping range of emotions apparent to a reader of the text does not even begin to address the repertory of practical training needed by an actor to perform them. The following is but one such challenge to the actor: how, indeed, does one invent a subtext for Bessie sufficiently deep and complex to support her turn to song at the moment of her death?

It would be tempting to conclude, for these reasons, first, that O'Casey's dramaturgy in *The Plough and the Stars* manifests a greater immediacy of event than that in evidence in the companion scene from the earlier play, and, second, that this shift toward a fully mimetic depiction of character and action represents the achievement of a more mature (or at least more technically proficient) dramatist, one who is more secure in his craft and who demands more from his actors. Certainly in performance the death of Bessie is both arresting and deeply moving. But any further argument for the relative inefficacy of the earlier narrative-based dramaturgy, however, would overlook two essential points about what happens in *The Shadow of a Gunman* as a consequence of O'Casey's transformation of enactment into narrative. There is no doubt that even in this early work O'Casey provides the audience with sufficient verbal and auditory information for them to form an imaginative picture of events they cannot see. Indeed, their attention is focused almost equally on and off stage. Before Mrs. Grigson rushes on stage to deliver her narrative, there are explosions and shots heard from outside the house, Seumas Shields and Donal Davoren cower in fright, and Adolphus Grigson dashes from the lounge to the relative safety of the kitchen. Eventually the racket offstage gives way to what O'Casey calls a "peculiar and ominous stillness" (O'Casey 129), and the audience hears snatches of conversation coming from the streets: "*Questions are heard being asked: 'Who was it was killed?' 'Where was she shot?' which are answered by: 'Minnie Powell'; ' She went to jump off the lorry an' she was shot'; 'She's not dead , is she?'; 'They say she's dead—shot through the buzzum!'*" (O'Casey, *The Shadow of a Gunman* 129).

Thus by the time Mrs. Grigson returns to deliver her narrative, the audience has already been given an instructional context for seeing the actions that they subsequently hear described. When Mrs. Grigson speaks, however, she does not by any means give a sophisticated account of offstage events; this is owing in part to her hysteria, but also to what comes across in the text as the limitations of her cultural background and education. O'Casey describes her as "*one of the cave dwellers of Dublin, living as she does in a tenement kitchen, to which only an occasional sick beam of sunlight filters through a grating in the yard* (O'Casey 113). In the first place, therefore, and in contrast to the richly informative method of Euripides' messenger, Mrs. Grigson's story appears on the face of it to be "badly" narrated. That is to say, it seems to be told by someone who is not fully in control of her story. This is an impression that derives not mainly from the narrator's excitement (a highly emotional state often characterizes Euripides' messengers), but more importantly from Mrs. Grigson's habit of complicating her oral delivery by addressing it to different audiences, first to Shields and Davoren, two other residents of the tenement, then by speaking apostrophes to the dead Minnie herself: "Oh, Mr. Davoren, isn't it terrible, isn't it terrible!" and "Poor little Minnie, poor little Minnie Powell, to think of you full of a life a few minutes ago, an' now she's dead!" (O'Casey 129–30).

The structural turbulence of Mrs. Grigson's narrative is an important part of O'Casey's dramaturgy, and her account of Minnie's death is further complicated by an overlapping of conflicting points of view. This is especially apparent in her final comment on the tragedy, which, within a single sentence (as quoted above) shifts from a direct address to Minnie to the more conventional narrative syntax: "an' now she's dead!" In translating experience into story, as Mrs. Grigson's language shifts back and forth, the stable relationship between voice (who is speaking?) and point of view (from where do we perceive what is being spoken?) becomes less distinct. The "message" of such a narrative stance, one might say, is as much in the telling of the story as the tale itself. One might conclude that when it comes to the representation of violence, narrative in *The Shadow of a Gunman* functions not just to provide a stimulus to the mind's eye but also to make a substantive point with respect to what is being told.

Two other modern plays that make extensive use of retrospective narratives in place of enactment are Tennessee Williams' *The Glass Menagerie* (1944) and Edward Albee's *The Zoo Story* (1959). The latter is a short conversation play involving two men who meet one afternoon on a park bench in New York. One, Peter, is a middle-aged, middle-class executive; the other, Jerry, is a kind of vagrant philosopher. He lives alone in a rooming house where his most meaningful social interaction is his daily confrontation with the landlady's dog. He is so desperately lonely that he picks a fight with Peter, whom he provokes into stabbing him, achieving by that act of suicidal sacrifice a bizarre kind of genuine human contact.

Formally speaking, the most interesting and novel element of the play is not its climax in a scene of onstage violence but two of Jerry's narratives, both involving offstage events. The first, the mysterious "zoo story" of the title, proves to be an instance of bait-and-switch. Never actually told, the promised story is referenced frequently enough to form a kind of organizing leit motif. The second story is told in a compelling scene that occurs when Jerry tells Peter a long, rambling narrative he calls "The Story of Jerry and the Dog." Jerry tells how he first attempted to win over with kindness the belligerent watchdog that guards the hallway that he must pass by each day on his way to and from his room. When love and kindness fail to change the dog's behavior, Jerry tries to kill it. The dog nearly dies, and from then on it simply avoids Jerry altogether; the animal shows him no responsiveness whatsoever, neither affection nor aggression, a situation Jerry finds in some ways even more distressing than when it clearly seemed hostile.

Like the Euripidean messenger speech, Jerry's monologue weaves together an elaborate texture of description, affect, and interpretation. It was perhaps this speech, surely one of the most famous sustained monologues of the modern theatre (in performance it can take up almost a third of the hour-long drama), that caused Thornton Wilder to criticize the relative inefficacy of Albee's play as theatre: "I don't think it would play half as well as it reads," he wrote in a letter to Albee, who had sent Wilder the manuscript for his review. "The men," Wilder continued, "the concrete men there—would get in the way" (www.http://nytimes.com.books/99/08/15/specials/albee-gussow. html). Albee himself seemed aware of the risk that so lengthy a monologue might become dull and "untheatrical"; he specifies in an editorial note that Jerry's speech should be done with a great deal of action to achieve a hypnotic effect on Peter as well as on the audience. Some of these gestures are embedded within Jerry's language, and, for the most part, they are the same kinds of mimetic acts that accompany the performance of conventional messenger speeches. For example, when Jerry is describing the dog's color, he mentions a "gray-yellow-white color...when he bares his fangs. Like this: Grrrrrr" (Albee 30). In this instance, the attempt to mime the actions of the absent dog, even to the extent of speaking the dog's words, as it were, has a close counterpart in the tendency of the classical messenger to make the characters in his narrative come alive by quoting them directly or imitating their gestures (de Jong 138). Similarly, when Jerry recounts his conversation with the butcher who sells him hamburger to give to the dog, he repeats the actual conversation, his diegesis again slipping into mimesis: "so I said, a little too loud, I'm afraid, and too formally: YES, A BITE FOR MY PUSSY-CAT" (Albee 32). But the degree to which the narrative is enhanced by mimetic interpolations is for the most part left open: Albee himself suggests only a few specific actions, in the belief that the director and the actor playing Jerry might best work out for themselves the mechanics of performing the speech (Albee 29).

The Zoo Story was given its premier at the Schiller Theater on September 28, 1959, where *Die Zoo-Geschichte* formed part of a double bill along with Beckett's *Krapp's Last Tape*, a play with which it shares a number of basic dramaturgical similarities. Certainly both works, as their titles suggest, are monologic rather than conventionally dialogic; they are essentially exercises in staging narrative discourse, in dramatizing the telling of stories and/or listening to them. Beckett's play makes use of a tape recorder to present "Krapp" at different times in his life. Krapp at age 69, whom we see on stage, listens off and on to a taped narrative made when he was thirty years younger; during those times when he is not listening passively to the earlier recording, Krapp attempts to make a contemporary tape, but this effort amounts to little more than an old man's sour commentary on the fool he used to be. We do not see this younger man, of course, but we nevertheless form a distinct mental impression of him based on the quality of his voice, which is quite different from that of the man who appears on stage; according to Beckett's stage direction, the tape produces a "strong voice, rather pompous, clearly Krapp's at a much earlier time."

Language in Albee's play is similarly monologic. Even though the play requires the physical presence of two characters, it is essentially the impulse to narrate, to tell stories, that motivates Jerry to speak. Jerry is both eyewitness and philosopher: "You're full of stories, aren't you?" says Peter. It is perhaps not particularly surprising that Albee should choose narrative rather than enactment in presenting the encounters of Jerry and the dog; animals on stage are notorious for their inability to conform to the roles, so to speak, that have been scripted for them. But the chief effect of Jerry's rant about his landlady's dog is to highlight the tension between narrative and enactment that is contained within the play, a tension that Albee, like Beckett, attempts to exploit both for formal as well as affective ends. Peter's role as a responder to Jerry's monologue is minimal; a few stage directions indicate that he occasionally displays emotional reactions to Jerry's discourse ("Peter winces"; "Peter raises a hand in protest"; "Peter indicates his increasing displeasure and slowly growing antagonism"; and so on). But for the most part Peter's role in the scene is that of a passive listener whose visibility (like the sight of Krapp bent over his tape recorder) is necessary in order to enact on stage what might be called "a story-telling event" (Fleischman 121).

In *The Glass Menagerie*, on the other hand, Williams creates a double role for the central character, Tom, who functions both as a narrator who is external to the play and as a character in his own right. *The Glass Menagerie* is constructed as a memory play. Its setting, according to Williams, is expressly nonrealistic, and the narrator, accordingly, is "an undisguised convention of the play who takes whatever license with dramatic convention is convenient to his purposes" (Williams 22). Thus Williams' play offers a kind of metatheatrical commentary on the narrator as a theatrical institution. Insofar as he is capable of addressing spectators directly, Tom provides a diegetic scaffolding for the play; whenever he breaks frame with his character or steps outside the mimesis, his remarks to the audience have a conspicuousness that reminds them openly of their engagement with a theatrical performance. "I am the opposite of a stage magician," Tom says in his introductory speech: "He gives you illusion that has the appearance of truth. I give you truth in the pleasant disguise of illusion" (Williams 22).

Because the narrator of *The Glass Menagerie* regularly reminds spectators of their active involvement with a theatrical event, he also provides Williams a supplemental cognitive frame for dramatic action whenever he steps intentionally out of the represented present to preface events by giving background information about characters or situations. At the beginning of Scene 6, for example, Tom addresses the audience to introduce them to Jim, a potential suitor for Laura:

> And so the following evening I brought Jim home to dinner. I had known Jim slightly in high school. In high school Jim was a hero. He had tremendous Irish good nature and vitality with the scrubbed and polished look of white chinaware. He seemed to move in a continual spotlight. He was a star in basketball, captain of the debating club, president of the senior class and the glee club and he sang the male lead in the annual light operas. He was always running or bounding, never just walking. He seemed always at the point of defeating the law of gravity. He was shooting with such velocity through his adolescence that you would logically expect him to arrive at nothing short of the White House by the time he was thirty. But Jim apparently ran into more interference after his graduation from Soldan. His speed had definitely slowed. Six years after he left high school he was holding a job that wasn't much better than mine. (Williams 68)

This is information that for the most part could have been delivered by way of expository dialogue, and there is no reason to think that Williams avoided doing so because of technical or thematic reasons. The problem of how to make the prior history of a given character known to the audience is more difficult, to be sure, for playwrights whose plots are not based on history or myth, but even so it is common for authors to replace narrative exposition in favor of animated dialogue or (at least) a conversational exchange of information. A writer concerned to maintain the illusion of a "fourth wall" between stage and audience would doubtless construct the scene so as to have Tom and his mother engage in a fond reminiscence along these lines about Jim:

> *Tom would say:* In high school, Jim was a hero.
> *And his mother would reply:* He had tremendous Irish good nature and vitality. He always reminded me of white chinaware.

And so on for the duration of the scene.

Granted, it's simpler just to sketch Jim by way of a narrative summary, but Tom's speech in its published form accomplishes more than simple exposition. When Tom steps outside the represented space of the play to speak directly to the audience from an adjacent but non-illusionistic location, his remarks tend in a peculiar way to stress the provisional nature of the subsequent action in which he is one of the principal figures. This speech announces the contents of the representation to an audience; it tends in this way to hollow out the subsequent mimesis, making it susceptible of analysis and classification. Especially in conjunction with the screen legends that are projected from time to time against the set—themselves a diegetic element analogous to Brecht's theatrical *Merkmale* (Jameson 44)—Tom's speech converts the ensuing scene into a mode of quotation. These narrative summaries ("The high-school hero"; "The Clerk"; "The accent of a coming foot"), like Tom, cause a tension between mimetic and diegetic spaces. Tom's situation is somewhat different, therefore, from that of the narrators in Albee's or O'Casey's plays, in that in speaking his narratives he makes use both of scenic and extrascenic space (Rehm 21). Even when he is not speaking directly to the audience, when he is engaged in conversation with other characters from within the representation, his narratives tend to be conspicuous as narrative in a way that those of Jerry or Mrs. Grigson are not. Speaking from a stance more fully within the diegetic mode, Tom is always in a position to narrate events but not so much to "represent" or to "perform" them.

Caryl Churchill and Brian Friel offer more recent examples of twentieth-century dramatists who purposively substitute narrative for direct scenic enactment. Friel's *The Faith Healer* consists of four lengthy narratives spoken by three different characters, Frank, Grace, and Teddy. Frank is the eponymous faith healer, Grace his long-time mistress and wife, and Teddy his Cockney manager. The play covers the events of many years in the lives of the trio as they travel back and forth across rural England, Scotland, and Wales, offering miracle cures to whoever is willing to believe in Frank's healing powers. The three characters never interact; in fact, they never appear together on stage, instead coming on serially to deliver their respective narratives. Each tells more or less the same basic story about their life on the road, and the same events regularly recur, each time told from a different perspective: the death of parents, Grace's stillborn child, Frank's miraculous curing of a man with a distorted finger. Grace and Teddy each speak one monologue, while Frank speaks two (the first and last of the play's four parts). Spaced throughout the four narratives are recitations of the names of the towns where they have played their act these many years. The names are spoken as lists, recited without context or affect, yet in performance they sound mellifluous, as enchantingly lovely as if they were poetry:

> Aberarder, Aberayron,
> Llangranog, Llangurig,
> Albergorlech, Abergynolwyn,
> Llandefeilog, Llanerchymedd,
> Aberhosan, Aberporth. (Friel 331–32)

Like most of Friel's works, *The Faith Healer* is biased generally toward conventional mimesis. Each of the characters speaks from within a naturalistic mise-en-scène. Friel specifies an auditorium with row seats and a large poster advertisement ("The Fantastic Francis Hardy / Faith Healer / One Night Only") for some of the scenes, while others take place in a generic room with a chair, a table, and a few props: cigarettes, an ashtray, bottles of whiskey or beer, some glasses. Yet clearly Friel structures his play so as to minimize any sense of direct scenic enactment—the sets are more like backdrops than illusionistic locations—and the power of *The Faith Healer* as theatre depends, therefore, mainly on the peculiarly mesmerizing effect of the spoken narratives. Alone on stage when they speak, Frank, Grace, and Teddy necessarily orient their narratives outward toward the audience; they are not speaking soliloquies or even interior monologues so much as simply telling stories to an audience, and it was this relatively "undramatic" dramaturgy of telling that consistently provoked reviewers' criticisms. (The phrase "theater of the mind" was commonly used by reviewers in describing Friel's play.) Clive Barnes, for example, expressed some doubt whether or not Friel's work was actually a play, choosing instead to call it "a theater experience," while Herbert Mitgang, writing for the *New York Times Theater Review* (November 12, 1983), commented that "you cannot help wondering and missing what might have happened if they [Frank, Grace, and Teddy] actually confronted each other in dialogue instead of monologue" (www.http://theater2.nytimes.com/mem/theater/treview/html?res=9402E2D91539F931A25752C1A965948260).

The patterns and rhythms of narrative completely dominate *The Faith Healer*. Each of the characters is given a number of actions to perform while speaking—removing an overcoat, smoking, drinking a glass of beer—but these actions are kept to a minimum and do not seem intended to create the illusion that we are seeing these characters as if through an invisible fourth wall. Their narratives—because they are not spoken with the ironic self-consciousness that marked Tom's performance in *The Glass Menagerie*—manifest more completely a storytelling theatre as opposed to a theatre that is conventionally "theatrical". Here the emphasis is shifted away from a mimesis that verbalizes events as one experiences or observes them toward a post-hoc verbalization that is presented as memory; Frank says that

> When we started out—oh, years and years ago—we used to have Francis Hardy, Seventh Son of a Seventh Son across the top. But it made the poster too expensive and Teddy persuaded me to settle for the modest "fantastic." It was a favourite word of his and maybe in this case he employed it with accuracy. As for the Seventh Son—that was a lie. I was in fact the only child of elderly parents, Jack and Mary Hardy, born in the village of Kilmeedy in County Limerick where my father was sergeant of the guards. But that's another story.... (Friel 332–33)

Churchill's *Mad Forest*, in contrast, is structured according to two kinds of perceptual acts, one dramatic and immediate, one retrospective and narratorial. Three months after the capture and execution of the Romanian dictator Nicolae Ceaușescu in the winter of 1989–90, Churchill and a troupe of actors from the London Central School of Speech and Drama traveled to Bucharest with the intention of creating a piece of theatre about current events in Romania. The first and third parts of the drama—called, respectively, "Lucia's Wedding" and "Florina's Wedding"—are conventional enactments of contemporary, private events and conversations in the lives of two fictional Romanian families (though the final act extends the range of characters to include a talking dog and a vampire). In these acts, the great, sweeping events of social and political change are deliberately moved into the background (as they were in Hauptmann's play about the Silesian weavers) in order to show "history" as it is lived by common working people. But the second part of the play, entitled, simply, "December," presents an entirely different mode of theatrical representation, one in which actual public events of the Romanian revolution are not sensorially present but are instead depicted by numerous, different acts of narrative picturing. As if to underscore the formal differences between these two distinct kinds of dramaturgy, telling and showing, Churchill replaces the characters of the first act with an entirely different cast: "None of the characters in this section," she writes in a stage direction, "are the characters in the play that began in part 1" (Churchill 29).

This replacement of drama by narrative speeches is closely related to another of Churchill's familiar dramaturgical features, the practice of "doubling," or having individual actors play more than a single role. Having actors take on multiple characters within a single play has long been part of theatre history—doubling was common in Greek tragedy, for example, where playwrights were limited to three actors—but twentieth-century playwrights such as Churchill often make virtue out of what was once a necessity. Having the same actor play more than one role can be similar to cross-dressing, insofar as both practices make visible the lack of "fit" between actor and role and so tend to throw into relief the idea that human identity is partly a theatrical performance. The result is first of all a kind of localized estrangement effect with respect to the perception of character. "Character" then becomes more an exterior quality or sociopolitical phenomenon, in much the way Brecht had theorized; for this reason, similarly, in *Cloud Nine*, Churchill requested the part of a little girl to be played by a grown man. In a more general sense, doubling can also work to undermine mimesis in that it subjects the representation of "real-time" events to a conspicuous fragmentation or autonomization (Jameson 43). Doubling roles precipitates out the mimesis into its elemental component parts; it slows down the representation, so to speak, so that the actor's taking on character becomes a kind of fledgling narrative in its own right.

The events that are recounted by the various characters in Part II of *Mad Forest* are not made available for seeing; it is understood that they have already taken place and can be reconstructed only by means of narrative. No attempt is made at scenic enactment in this part of Churchill's play, nor, for that matter, is there any attempt made to place the various characters in any credible representational context. The stage is bare, and each of the twelve characters is identified only by type, as, for example, "painter," "girl student," "translator," "bulldozer driver," and so on. These characters stand before the audience, and one after the other they tell brief stories concerning the events they saw taking place during the revolution. The men and women who speak these different parts before the audience are not really "acting," at least in the conventional sense of taking on an identity different from their own. They appear "in character" only to the extent of adapting Romanian accents, but they do not interact with each other even though they share the same physical space. Rather, as Churchill specifies in a stage direction, "each behaves as if the others are not there and each is the only one telling what happened" (Churchill 29).

The difference between this act and the first and third acts of *Mad Forest* can be explained by analogy with the classical distinction between showing and telling as modes of poetic representation, between dramatizing characters and events as if they were present, on the one hand, and, on the other, representing them only retrospectively in words, as reported history. In comparison with the different enactments of Acts I and III, the speeches in Act II do not necessarily lack vitality and vivacity; they are contemporary kinds of "messenger speeches," and, like those ancient narratives, they have a highly complex sensory potential. But given that the narratives of Act II formally contradict the mimetically represented world of Acts I and III, it seems clear that Churchill's intent with them is to demonstrate the potential of mere words to supply the enargeia, or vividness, that makes hearing a credible alternative to sight. For example, Student I says, "At six in the morning there is new tar on the road but I see blood and something that is a piece of skin. Someone puts down a white cloth on the blood and peoples throw money, flowers, candles, that is the beginning of the shrines" (Churchill 35).

One might call this account a variety of "thick" description; in referring to a number of particular colors and objects the speaker gives spectators a formula or recipe for picturing a scene that they cannot actually see. But the imagined vividness of this scene depends also on the speaker's apparent distance from the real dramatic action he describes. He and the other narrators in this part of the play seem to function not at all like mimetic characters but more like historians or chroniclers. The narratives exemplify the relation of unmediated event, which is the conceptual basis of drama, to the condition of retrospective account or "story," which is the condition of narrative. "Drama," says Susanne Langer, "moves not toward the present, as narrative does, but toward something beyond.... As literature creates a virtual past, drama creates a virtual future. The literary mode is the mode of Memory" (Langer 307). Another way to put this would be to say that the narrative voices in Part II of *Mad Forest* provide for a secondary imaginative response to the play, a response that is evocative, memorable, at times uncanny. As a set of marginal glosses may be said to transcend the occasion of the text (Lipking 611), these narratives transcend the occasion of the mimesis in which they happen to find themselves.

A final variety of "messenger narrative" in twentieth-century drama would be the performance piece by Anna Deavere Smith in *Fires in the Mirror* (1992), a play in which narrative assumes not a supplemental function but the central one. In fact, Smith's drama is rendered entirely in narrative, entirely (to use Langer's phrase) in the mode of memory. Smith developed the play in response to the rioting and killing that took place in the Crown Heights neighborhood of Brooklyn in August 1991, after a car carrying Hasidic Jews ran a red light and accidentally struck two young black children. One of them, a seven-year-old Guyanese boy named Gavin Cato, was killed. Immediately after the accident, a crowd of African Americans gathered and grew swiftly belligerent. One ambulance, a Hasidic emergency vehicle, arrived within three minutes of the incident; the police and a city ambulance were on the scene only a short time later. All accounts of the early events agreed that the driver of the automobile involved in the accident, Yosef Lifsh, was being beaten by an angry crowd by the time the police arrived. What happened next is only slightly less clear: police reports state that the officers on the scene ordered the crew of the Hasidic ambulance to tend the Jewish men and leave the scene immediately so as not to antagonize the crowd with their presence. This they did, but their departure was apparently misunderstood. Rumors soon spread that the Hasidic ambulance personnel had treated the injured Jewish men while ignoring the children, and these rumors caused the deep and long-simmering anti-Jewish sentiment within the community to boil over into ugly anti-Semitic rioting that lasted three days. Late that evening a large group of black youths assaulted Yankel Rosenbaum, a visiting Australian scholar who had the misfortune to be in the wrong place at the wrong time; the gang beat Rosenbaum until his skull fractured, but what most likely killed him were stab wounds inflicted by one of them, a sixteen-year-old named Lemrick Nelson.

Smith dramatizes nothing of those events themselves, which remain unseen; instead, history is rendered by means of a sequence of monologues spoken by a great number of different characters with varying perspectives on the Crown Heights tragedy. The play is structured as a version of journalistic oral history (Smith freely acknowledges her indebtedness to the methods of Studs Terkel); it consists of Smith's edited transcripts of various interviews about the riot that she conducted with more than two dozen people. Smith gives additional shape to the various monologues by editing them and arranging them on the page more like poetry than prose. She also provides a measure of enactment for each monologue by giving details about the speakers' situations and physical characteristics, and she makes occasional use of props or bits of costume. But the play as written is in no sense conventionally mimetic.

As Smith originally conceived the piece, she herself performs all the roles, further marking the dramatic components of character and action as inseparable from their transmutation into narrative. The play has sometimes been performed with larger casts—there has been at least one full-cast professional performance by the Mixed Blood Theatre in Minneapolis (Worthen, *Print and the Poetics of Modern Drama* 107), while a performance in New York by City Theatre in 1995 used one white and one black actor, supposedly to underscore the hostile tensions between the races. Still, from a dramaturgical point of view, Smith's solo performance of all twenty-six roles offers at least one distinct advantage over a more naturalistic or ensemble style of performance. Just as Churchill used doubling to show that character might refer to qualities that one puts on or performs, rather than any intrinsic aspects of identity or personality, a solo performance of *Fires in the Mirror* allows for a much different kind of analytical frame. In this case the result is not so much to make race the sole or even the central issue of the play, by focusing on the social and cultural underpinnings of character, as to bring about the wholesale conversion of mimesis into a kind of "saying." The utter simplicity of a solo performance engenders a distinctly epic or narrative effect, like a series of still-life portraits set side by side.

A number of the speakers, such as Carmel Cato and Norman Rosenbaum, the father of the dead child and the brother of the murdered Jewish scholar respectively, are conventional narrators from a dramaturgical point of view. They are either participants in events or witnesses to them; accordingly, they give their varying accounts of events in much the same manner as the messengers of ancient Greek tragedy. But some of the other speakers do not seem to be eyewitnesses at all. In fact, they scarcely could offer any such testimony, since many of the people who speak did not see any of the events; their knowledge of them comes from the news media or word of mouth, and in one instance Smith's character's words are transcribed from interviews conducted by Smith two years before the Crown Heights rampage occurred. Thus the section entitled "Rhythm" contains a rambling narrative about one of Smith's former students, Monique "Big Mo" Matthews, on the subject of "performance":

> An she say, "This is for the fellas,"
> and she took off all her clothes and she had on a leotard
> that had all cuts and stuff in it,
> and she started doin' it on the floor.
> They were like
> "Go, girl!"
> People like, "That look really stink."
> But that's what a lot of female rappers do—
> like to try to get off,
> they sell they body or pimp they body
> to, um, get play" (Smith 35)

This interview, according to Smith's notes, took place in the spring of 1989 when she was a resident scholar at UCLA (Smith 35). Even though it has a measure of relevance with respect to Smith's generalized theme of race and identity, the effect of Big Mo's narrative is neither explicitly historical nor conventionally ecphrastic. Rather the monologue is poised halfway between the art of verbal scene-painting, where the original actions are made visible through retrospect in language, and the filmic technique of montage, where (as Sergei Eisenstein theorized) the juxtaposition of contrasting shots could create meaning apart from anything inherent in either of the shots themselves. Smith's play is structured according to a principle of such ironic contrasts, so that any one given narrative is set in the context of others that echo it or contradict it, either directly or tangentially. Thus, for example, a character identified as Aaron M. Bernstein, a middle-aged Cambridge physicist, talks about the refractive capabilities and limits of mirrors, while the subsequent narrative is attributed to an unnamed "teen-age Black girl of Haitian descent" who speaks while looking in a mirror (Smith 16).

The monologue that opens the play, though based on a conversation that took place when the speaker was fully aware of the Crown Heights riot, also makes use of a similar dramaturgy of circumlocution in that the event that forms the play's thematic center is not explicitly named. This introductory narrative was adapted from an interview Smith conducted by telephone with the playwright Ntozake Shange. Shange's speech is about identity, a subject of concern that immediately links it with the events of Smith's play mentioned in the subtitle: *Crown

Heights, Brooklyn and Other Identities." "You don't know," Shange says, "what you're giving if you don't know what you have and you don't know what you're taking if you don't know what's yours and what's somebody else's" (Smith 4). But the rhetorical strategy of Shange's conversation, like that of "Big Mo," also turns aside from a straight-forward history of recent events, never actually naming the things it purports to address.

A number of other sections of *Fires in the Mirror* feature familiar public personages such as Al Sharpton, Leonard Jeffries (then a contro-versial professor of African American Studies at the City University of New York), and Letty Cottin Pogrebin (a founding editor of *Ms.* mag-azine); these people address the Crown Heights tragedy more directly, even though none of them were witnesses to the action in the way of traditional messengers. They are real people—that is, they represent actual, living individuals, recognizable by most people in the audience, and the words that Smith performs are their own responses to what they heard, saw on the news, or read about the riot. But their testimony, because they engage with events only from a distance, has definite limi-tations with respect to providing the audience with a "picture" of what happened on August 19, 1991; Pogrebin's commentary is an example:

> I think it's about rank frustration and the old story
> that you pick a scapegoat
> that's much more, I mean Jews and Blacks,
> that's manageable....
> To get a headline,
> to get on the evening news,
> you have to attack a Jew.
> Otherwise you're ignored.
> And it's a shame.
> We all play into it. (Smith 50–51)

When we test many of the speeches in *Fires in the Mirror* against the standards and qualities we expect to find in conventional eyewitness narratives, they differ in a number of respects. It is not simply that they swerve from history. The speeches of the characters almost all have a substantial basis in fact. Yet none of these speakers directly addresses the real subject of Smith's play, and it is precisely this avoidance that gives their testimony its compelling power. Certainly one could make a case here for the persistence of modernist poetics in the drama of the late twentieth century, because there is in Smith's haunting performance piece something of the ambiguous, democratic chaos of voices heard in the poetry of T. S. Eliot or the narratives of Gertrude Stein and Samuel Beckett. Perhaps, also because so many of the speeches are so obviously not the direct commentary of eyewitnesses, the voices seem to take on elliptical, almost sibylline qualities. Consider, in this regard, the speech of an anonymous Lubavitcher woman, the second speech in the play:

Well,
it was um,
getting toward the end of Shabbas,
like around five in the afternoon,
and it was summertime
and sunset isn't until about eight, nine o'clock,
so there were still quite a few hours left to go
and my baby had been playing with the knobs on the stereo system
then all of a sudden he pushed the button—
the on button—
and all of a sudden came blaring out,
at full volume,
sort of like a half station
of polka music.
But just like with the static,
it was blaring, blaring
and we can't turn off,
we can't turn off electrical,
you know electricity, on Shabbas....
so I went outside
and I saw
a little
boy in the neighborhood
who I didn't know and didn't know me—
not Jewish, he was black and he wasn't wearing a yarmulke
because you can't—
so I went up to him and I said to him
that my radio is on really loud and I can't turn it off,
could he help me,
so he looked at me a little crazy like,
Well?
And I said I don't know what to do,
so he said okay,
so he followed me into the house
and he hears this music on so loud
and so unpleasant

and so
he goes over to the
stereo
and he says, "You see this little button here
that says on and off?
Push that in and that turns it off."
And I just sort of stood there looking kind of dumb
and then he went and pushed it,
and we laughed that he probably thought:
And people say Jewish people are really smart and they don't know
how to turn off their radios. (Smith 7–8)

The woman's narrative is in certain ways more ostensive than would be a fully mimetic dramatization of the conversation in its proper chronological relationship with the rest of the events that occurred in Crown Heights in late August. A mimesis of the encounter between the woman and the boy could certainly equal or exceed the narrative in portraying the incident with accuracy and humor, but such a dramaturgy of "presentness" would likely be more tonally neutral and innocent in comparison with the narrated version. This is not to say that a skillful enactment of the incident could not project how generally fraught are the interactions between people whose cultures are mutually ignorant of one another. But any such foreshadowing would necessarily present itself almost accidentally and incidentally. It would be more like a by-product of the performance rather than a central effect, owing for the most part to the tendency of audiences to project their own ideas for formal and thematic cohesiveness onto an assemblage of scenes that are not explicitly linked but merely juxtaposed. Something like this occurs frequently in film, where the technique of montage is used to create meaning independent of and distinct from anything actually recorded in the individual shots.

In the case of Smith's play, however, the comments of the Lubavitcher woman, because they are spoken in the form of narrative, have the perspective of hindsight; they are understood, therefore, to be causally suggestive. At the moment she speaks, the woman adopts the point of view of someone who at the moment she is recounting the story already knows its outcome. When, for example, she speaks about the effects of the radio's noise, she describes "a young boy that was visiting us, / he was going nuts already, he said, / it was giving him such a headache could we do something about it." In this passage the woman reports the activity both from an objective point of view, telling her story in the historical mode as a completed past event, and also from the point of view of the boy's experience as he is experiencing it, by employing what is called free indirect discourse in which the words or thoughts of the character are subsumed into the discourse of the narrator: "giving him such a headache could we do something about it." What is captured in this excerpt is the way the mimesis of the boy's immediate speech is made to correspond to the narrator's discourse; that is to say, the events are recounted from within a narrative stance that combines two modes of telling: mimesis through diegesis.

The stage narrator stands before us as an eyewitness to important events, but a narrator's report is not just a set of words that help listeners picture those unseen sights. That is only one of its important effects. The narratives are also a blueprint for experiencing those events in a particular instructional way. The classical messenger typically speaks, as James Barrett has argued, with a narrative voice that closely resembles that of epic and, therefore, endows his words with a high degree of authority (Barrett xvii). The classical stage narrator thus performs a kind of autopsy on an unseen event, presenting that analysis for the audience. But *Fires in the Mirror* offers not one but more than two dozen narrators, each with an equal claim, in principle, to authority and truth. In the absence of a single unifying narrative stance or point of view, the many different voices, the majority of whom give only partial or contradictory pictures of events, add up to no single coherent account. Instead the multitude of voices offers the audience a kind of puzzle, as elements of hearsay, eyewitness accounts, private conversations, and political commentary all coexist on the same stage and compete for the audience's assent. Therefore, although Smith's dramaturgy on the whole depends on some of the familiar qualities of the classical messenger, there is a sharp distinction between Smith's method and the narrative strategies of more conventional representations of offstage action. In *Fires in the Mirror*, the multiplicity of narrative voices renders all speech equally authoritative—or in the context of the play, equally suspect. One follows along with no more than fragmentary clues into events and into the motives of the people who experienced them or who tell them. It is as if something vital were cut from the heart of history. This is, of course, a powerful formal device, because it invites us alternatively to take sides in choosing which version of history to believe, for example, either against the Jews, who appear regularly to use Holocaust stories as a means of claiming the moral high ground, or against the African Americans who react so mindlessly to what they wrongly perceive as Jewish arrogance. Such is the power of Smith's montage to frame the contemporary public debate surrounding race relations in the United States; in the absence of any dominant narrative voice, Smith's dramaturgy shows the way that discourse turns event into history and history into legend. Each of the narratives in *Fires in the Mirror* is tragically limited—which is, of course, Smith's intention. It's hard to imagine a play structured like *Fires in the Mirror* being staged successfully except before audiences of the twentieth and twenty-first centuries—that is, audiences with a Romantic bias for fragments or innuendoes, and, especially, audiences who believe that all truth is relative and that history is nothing more than a tale that changes according to the politics of its teller.

* * *

The deep moral ambivalence that comes of watching staged violence has often been the subject of speculative, even bewildered commentary: how can we explain, let alone justify, taking pleasure in watching representations of others' pain? It may be that in this respect the classical dramaturgy, which rendered violence by means of a report from an eyewitness, is close to an ideal ethical answer. In *Mad Forest* or in *Fires in the Mirror*, as in *Medea* or *Agamemnon*, the violence, brought forth retrospectively as narrative, is transformed into an object of cognition.

But what about the other great taboo subject of Western theatrical representation—sex? What if the unseen involves not violence but sexual acts? How have dramatists used the offstage and narrative to depict that? One usually thinks of the high-profile obscenity cases in the modernist era as involving novels (Lawrence's *The Rainbow* in 1915, for example, or Joyce's *Ulysses* in 1921). In fact, much of the early public discourse in the twentieth century about obscenity and censorship was aimed not at books but plays, where the material presence of the actor was assumed to result in an almost mesmerizing appeal for the audience. Often the debate was framed in terms of Plato's distinction between mimesis and diegesis, showing and telling; thus in a report published in 1909, a parliamentary committee on censorship in Britain recorded the testimony of the dramatist Sir William Gilbert (of Gilbert and Sullivan) that "[i]n a novel one may read, 'Eliza stripped off her dressing-gown and stepped into her bath,' without any harm, but...if that were presented on the stage it would be very shocking" (Peters 211).

In determining that to hear a narrated description of an erotic event was somehow less sensually arousing than to see it enacted, the committee members made the assumption, familiar in the history of Western theory on the arts, that visual signs in particular, because they are present to a beholder all at once (rather than singly and serially, as, for example, in a poet's narrative), tend to hold us in greater thrall. But there are important differences between the kinds of diegesis that dramatists use to frame the representations of violence and those involving sexual acts. Even in the twentieth century, where playwrights have greater liberty than ever before to stage erotic actions publicly and mimetically, there seems to be no general agreement regarding the extent to which the actors' mimetic representations of desire evoke responses from the audience that are indistinguishable from those normally reserved for people in real (i.e., nontheatrical) situations. This is why the film industry still maintains a form of censorship in its rating system and why parents try to restrict what their children watch by installing blocks in their television sets or computers. That watching actors on stage mime sex acts is itself sexually arousing is often said to be the position of prudes and reactionary zealots, but there are a number of reasoned contemporary arguments to support it. Perhaps what is needed for the visual arts, as David Freedberg suggests, "is that we reconsider the whole insistence in Western art theory—as, perhaps, in Western philosophy as a whole—on the radical disjunction between the reality of the art object and reality itself" (Freedberg 436). When an actor on a stage pretends to be a king, as Brecht and Plato well knew, an audience will respond or will be inclined to respond as if a king himself were actually present before them, and the only way to control the kind and degree of their engagement, therefore, is to cause them somehow to step back, to get a grip on their emotions and feelings, so to speak, in order to distinguish intellectually between reality and its aesthetic counterpart.

In contrast to offstage events involving violence or fantastic occurrences in which what is not seen is asserted retrospectively to be present through language and so partly transmuted by it, the theatre has tended historically to frame actions involving sexual activity not as retrospective narratives but as gaps or omissions. The invisible (but typically much anticipated) act becomes an event from which even narrative turns away. This suggests an intriguing variation in the way theatre has typically assigned diegetic space to violence, on the one hand, or, on the other, to sexual activity. The theatre's solution to the problem of representing erotic acts is often different from its representations of murders or battles; reporting sex is not done with the same methods or ends as the reporting of violence or death, and the conditions of its dramatization as an offstage event vary widely. In Renaissance plays, for example, as Celia Daileader has shown, the representations of offstage sexual acts tend to reflect the overall difficulties of staging erotic acts on an all-male stage. "[T]he prohibition against female actors," writes Daileader, "working in tandem with social decorum, encouraged the offstage placement of heterosexual intercourse; paradoxically, to make it remotely convincing, it would have to remain unseen" (Daileader 48). This seems to me a reasonable conclusion with respect to Renaissance dramatic performances in the public theatres, one that is certainly in keeping with Horace's admonition that the mimetic enactment of certain kinds of acts left him unbelieving. Better, for purposes of verisimilitude, to have these things told about rather than shown. But Daileader's summary begs the question of the extent to which local conventions themselves dictate what is perceived by an audience to be credible and what is not. When, in the late sixteenth century, a French theatre company that included women in its number toured London, the representing of female characters by female players was derided by spectators as being "unnatural."

In a modern theatrical context, on the other hand, where convention dictates there be a credible physical correspondence between actor and role, the casting of women in female roles allows for a relatively high degree of visibility when it comes to the depicting of heterosexual love. The drama of the twentieth century, in particular, offers playwrights almost unprecedented freedom in this respect. There are, of course, numerous post–World War II plays in which actors perform sexual acts or appear to perform sexual acts in full view of the audience; often these were created with express political or ideological purposes, as, for example, in the American underground theatre of the 1960s in Julian Beck's and Judith Malina's productions for the Living Theatre, or in the bizarre sexual images in works such as Jean-Claude van Itallie's *America Hurrah!* or Andy Warhol's *Pork*. Dramatists of the latter decades of the twentieth century such as Sarah Kane or the Austrian playwright Franz Xaver Kroetz feature in many of their plays fully mimetic enactments of various sexual acts, including masturbation as well as oral and genital sex. Kane's short career as playwright (she took her own life in 1999) began with *Blasted* (1995), a work that includes realistic depictions of frottage, oral and anal sex, rape, murder, and cannibalism; in Scene 2 of that play, for example, Cate, a young London woman, performs fellatio on Ian, a middle-aged Welshman, and in Scene 4, a soldier rapes Ian, first with his penis and then with the barrel of his revolver, and then sucks one of Ian's eyeballs out of its socket. Kane's next work, *Phaedra's Love* (1996), a rewriting of the ancient tale of Phaedra's illicit passion for her stepson Hippolytus, includes one scene in which Phaedra sucks on Hippolytus' penis until he ejaculates, and the play concludes with dramatizations of Theseus raping a townswoman and the cruel mutilation of Hippolytus' corpse.

All of Kane's works were dismissed by London audiences and reviewers as merely sensational when they were first staged, but more recent productions have been assessed much more sympathetically, no doubt in part because they are now inevitably seen through the lens of Kane's sadly brief life. In any event, the overall arc of Kane's short career seems to have taken the form of a swerve away from the brutal naturalism initially thought to represent the core of her dramaturgical style. This tendency seems to be present even in the early *Blasted*, whose final scenes, despite the visual horror of the actions they depict, move slightly away from absolute verisimilitude toward a more stylized or symbolic mode of representation. One can see clear evidence of this tendency in the final moments of the play, when Ian, having just eaten off the corpse of a baby, lies down in a hole so that only his head is visible. Kane's stage notes indicate first that Ian dies with relief as rain falls on him through a hole in the roof. Moments later, however, in a coup de théâtre reminiscent of some of the fantastic extravaganzas of the early surrealists, Ian's head continues to carry on a mundane conversation with Cate, who comes on stage carrying bread, sausage, and a bottle of gin:

> *Cate*: You're sitting under a hole.
> *Ian*: I know.
> *Cate*: Get wet.
> *Ian*: Aye.
> *Cate*: Stupid bastard. (Kane 60)

By the time of her last work, the enigmatic *4.48 Psychosis*, Kane seems to have all but abandoned mimesis as a dramatic strategy. The play's title refers to a time of day, 4:48 a.m., the hour when the night is darkest and when, therefore, the spirit senses most keenly the bleakness of existence. No individual characters are identified as being necessary to the play, and the work has been performed, variously, with different numbers of actors—two, three, sometimes more—dividing the speeches arbitrarily. Sometimes the play has even been done as a monologue.

If *4.48 Psychosis* includes no characters, neither is there setting or plot; the drama consists of a series of choric speeches apparently lacking in context or motive. The work begins with what seem to be fragments of a person's conversation with him- or herself:

> (*A very long silence.*)
> But you have friends.
> (*A long silence.*)
> You have a lot of friends.
> What do you offer your friends to make them so supportive?
> (*A long silence.*)
> What do you offer your friends to make them so supportive?
> (*A long silence.*)
> What do you offer?
> (*Silence.*). (Kane 205)

But the piece soon departs from any semblance of verisimilitude; in the printed text the words and sentences are arranged on the page not as components of conversation or discourse but more like choric responses, somewhat in the manner of a modernist poem. There are passages, for example, in which the speaker (or speakers) seems to channel global feelings of shame and guilt:

> I gassed the Jews, I killed the Kurds, I bombed the Arabs, I fucked small children while they begged for mercy, the killing fields are mine, everyone left the party because of me, I'll suck your fucking eyes out send them to your mother in a box and when I die I'm going to be reincarnated as your child only fifty times worse and as mad as all fuck I'm going to make your life a living fucking hell I REFUSE I REFUSE I REFUSE LOOK AWAY FROM ME[.] (Kane 227)

And there are passages that clearly reference contemporary medical discourse, in particular the language of clinical diagnosis and treatment:

> Sertraline, 50 mg. Insomnia worsened, severe anxiety, anorexia, (weight loss 17 kgs,) increase in suicidal thoughts, plans and intention. Discontinued following hospitalization. (Kane 223)

Certainly in comparison with her earlier works, *4.48 Psychosis* is much easier to stage in one respect because there are no scenes involving the depiction of violence, death, or sexual acts. ("It messed with my head," said Sarah Benson, who directed a recent performance of *Blasted*. "It was very disturbing. I was actually depressed" [*The New York Times*, November 6, 2008, sec. C5].) It is as if in constructing *4.48 Psychosis* Kane seems to have acknowledged the practical and emotional difficulties that her previous plays had presented to actors and audiences because of their dependence on mimetic enactments. To the extent that it is possible to draw general conclusions on the basis of such a limited body of work, one could say that with *4.48 Psychosis* Kane seems deliberately to have abandoned mimesis in favor of experimenting with a largely diegetic dramaturgy. Indeed Christopher Isherwood, writing a review of a production of *4.48 Psychose* (a French translation of the play performed in October 2005, at the Brooklyn Academy of Music's Harvey Theater), claimed that Kane's drama seemed to function as a "renunciation" of theatre (http://theater2.nytimes.com/2005/10/21/theater/reviews/21psych.html). In calling the play a "renunciation" of theatre, Isherwood could only have been referring to the manner in which Kane seems to have chosen not to visually represent characters, events, objects, or even a coherent and locatable space, replacing mimesis with a largely diegetic dramaturgy.

The dramas of Franz Xaver Kroetz, like those of Sarah Kane, in some ways represent a late twentieth-century return to naturalist or even verist principles and dramaturgical practices. Kroetz eschews what is perhaps the most basic theatrical convention, the premise that no matter what may be their education, social background, or material circumstances, all characters in a play are able to speak their minds, to make their thoughts and emotions accessible to an audience. Kroetz creates a dramaturgy for those who are largely unable to speak for or about themselves. His is a drama of the inarticulate, a theatre about people whose problems lie too deep for words. The majority of his plays feature scenes involving physical violence or sex acts. In *Stallerhof* (1971), for example, an aging farmhand named Sepp sits on a toilet and masturbates; it is the only action in the entire Scene 3. *Männersache* (*Men's Business*, 1973) depicts Martha, a butcher, and Otto, a construction worker, having sex on stage (in Scene 1 and again in Scene 7), and at the end of the play (Scene 8) they engage in a bizarre duel using Otto's rifle, taking turns shooting at one another. Martha fires first ("I'll start 'cause I'm the woman," she says), wounding Otto in the shoulder. Then Otto takes the rifle and shoots Martha. They shoot each other twice more; the final exchange takes place as follows:

> *Martha*: Right, there've got to be rules. (*shoots and hits*)
> *Otto*: (*tumbles back toward the wall*)
> *Martha*: (*puts the rifle on the floor, shoves it toward him with her foot*) There it is.
> *Otto*: Thanks. (*Aims, shoots, hits;* Martha *falls down face forward and remains lying there rigidly.* Otto *looks at her, comes closer.*) You give up? (Kroetz 99)

Whether explicit depictions of violence such as called for by Kane or Kroetz are ultimately helpful or hurtful to audiences who see them is a question about mimesis that has been debated for millennia, and debated without resolution. Audiences for a recent production of Kane's *Blasted*, for example, describe an experience in which the mental act of spectatorship becomes transformed into actual bodily sensations; they feel yanked about, shoved, or exhausted by the performance, yet they still queue up nightly to watch it (*The New York Times*, November 6, 2008, sec. C5). What is less problematic, however, are the responses actors have to performing in a work like this. For this one performance run, at least, the evidence suggests that actors had a hard time accommodating themselves to the mimesis without in some way detaching their physical actions from any affective state. Kane's play seems to require them to become immune to any moral context for the mimesis they are required to perform: "Once we got up on our feet in rehearsals," said one actor, "it just became the play, and it became the same as if we were doing another play—it's a breakfast scene, or it's an eyeball-sucking scene" (*The New York Times*, November 6, 2008, sec. C5).

I should say at this point that I am not interested here in discussing the huge subject of pornography, definitions of which are notoriously slippery if not impossible. Within the broad scope of the visual arts, writes Freedberg, "[w]hat is called pornographic remains wholly contextual; there can be no hope of deriving the term transcontextually or on anything remotely approaching a transcultural base" (Freedberg 356). Much the same principle holds true for theatre history, where dramas that have at times been called pornographic cannot be identified definitively either by their content or by the responses they typically evoke. As in the visual arts, pornography in the theatre, writes John Elsom, "is not a simple escalation process, from bad to worse.... [I]t is the search for an erotic image which is suitable on many levels—which may assuage masculine guilt and feminine shame, which reflects certain class relationships and which is sexually stimulating at the same time" (Elsom 239). Neither am I much concerned in this chapter with the question of whether looking in Western art is structured along gendered lines, though the evidence, so far as I am aware, suggests that in some ways it is (Freedberg 324–25). Rather at this point I simply want to conclude a discussion of retrospective narratives and offstage events by examining a few representative plays in which playwrights have dramatized sexual encounters by assigning them to a space somewhere beyond the mimesis.

Much of the problem of representing violence that occurs in nonmimetic space was solved both dramaturgically and ethically by substituting ecphrasis for the missing faculty of vision. But the formulations for pictorializing violence are not generally available when it comes to representing sex on stage with language. In fact, the difference between the ways these two kinds of unseen events are represented verbally is striking. One conventional feature of the "messenger speech" in ancient tragedy is its lucid, almost savory style. In *The Persians*, the herald reports that after the battle of Salamis "all who survived that expedition, / Like mackerel or some catch of fish, / Were stunned and slaughtered, boned with broken oars / And splintered wrecks" (Aeschylus 63). In *The Women of Trachis*, Heracles smashes Lichas against a rock by the sea, and the force of the impact, according to the messenger, "pressed the pale brains out through his hair, / and, split full on, skull and blood mixed and spread" (Sophocles 100). And in *The Bacchae*, an eyewitness reports how the crazed followers of Dionysos claw Pentheus' ribs clean of flesh and play ball with scraps of his body. So meticulously detailed are these accounts of wounding and mutilation that one has to conclude that the majority of Athenians expected during the course of tragic performances to be able to glut their imaginations on terrifying descriptions of at least one hapless person's agony.

The motive for recounting these events in narrative is twofold: first, to bring about an imaginative resurrection of the event, and next, at the same time, to make use of the retrospective vantage point of narrative to introduce a "shaped" or cognitive response to events rather than an immediate and largely emotional one (Macintosh 129). But it seems to be the case that representing sexual acts that take place offstage involves somewhat different techniques from those used to represent violence or

death. For when the imagined event is an erotic encounter, its substanti-
ation as an offstage event depends less on one or more characters' verbal
reconstruction than an ongoing dialectic between stage and audience
as to what is shown and what can be neither shown nor named. In contrast to the imaginative picturing of violence, picturing sex involves a
kind of visual sleight of hand and the frisson that always attends the
apparent transgression of culturally defined borders. "In the figurative
arts," says Mario Perniola,

> eroticism appears as a relationship between clothing and nudity.
> Therefore, it is conditioned on the possibility of movement—transit—
> from one state to the other. If either of these poles takes on a primary or
> essential significance to the exclusion of the other, then the possibility
> for this transit is sacrificed, and with it the conditions for eroticism.
> (Perniola 237)

It is possible, therefore, to acknowledge that in some cases a playwright might place sexual encounters offstage for reasons of practicality
(the Renaissance stage on which the parts of women were played by
adolescent boys would be an example) and yet still take advantage of
that visual absence in order to bring about a cultivated sense of arousal.
Strindberg's tragedy, *Miss Julie* (1888), is one such drama. The play is
based on the naturalist premise that sex, like death, is a great social
leveler. The plot centers on a one-time sexual encounter between Julie,
the daughter of a Swedish count, and Jean, who works in the household
as a valet. When, after the two have engaged in sex, it becomes apparent to Julie that she has no practical future as the wife or mistress of a
commoner, she decides to take her own life. Jean lives, but by the end
of the play he has lost all of his early charm and bravado, and, in comparison with Julie, who in dying attains a kind of tragic grandeur, Jean
is exposed as petty and craven.

In this as in some of his other plays, Strindberg depicts the relations
between men and women as inherently antagonistic and even hostile, a
biologically driven opposition that causes the sexes to hate each other
even as they find each other compellingly attractive. Thus Julie's tragedy, according to Strindberg, is due largely to forces beyond her knowledge or ability to control, especially "the festive mood of Midsummer
Eve, her father's absence, her monthly indisposition, her pre-occupation
with animals, the excitement of dancing, the magic of dusk, the strongly
aphrodisiac influence of flowers, and finally the chance that drives the

couple into a room alone—to which must be added the urgency of the excited man" (Strindberg 95).

Within the restrictive conventions of late nineteenth-century commercial theatre as they applied to the staging of erotic desire, Strindberg could not have represented his characters in a state of undress, much less attempted to dramatize actual sexual intercourse. Spectators must, therefore, complete his drama by imagining Jean and Julie having sex in a space that happens to lie just beyond view, in much the same way they are intended to complete with their imaginations the décor of the kitchen, only a portion of which, according to Strindberg's directions, is visible within the mise-en-scène. Spectators are meant to extrapolate the unseen portions of a kitchen stove and dinner table on the basis of those parts of the objects that are contained in the set:

> To the left is the corner of a large tiled range and part of its chimney-hood, to the right the end of the servants' dinner table with chairs beside it. (Strindberg 98)

Strindberg's method for representing sex, therefore, is consistent with the aesthetics of his décor, where those parts of objects that are visible are intended to stimulate viewers to fill in the rest that is veiled. Even narrative discussions of sexual encounters, such as the one that Julie and Jean have when they return to the stage, are veiled to the extent that their hiddenness is itself understood to be a message in its own right. Julie and Jean are seen to enter his room, and as soon as the door closes a group of peasants celebrating Midsummer's Eve come on stage, singing, drinking, and dancing. The point of the ballet-like interlude is first of all verisimilitude, to further the illusion that enough time passes for Jean and Julie to engage in sexual intercourse in the imagined space of the room just offstage. Yet the music and dancing Strindberg provides aren't there just to make the drama more credible in allowing for the passing of time for sex to occur. Dancing and music are presented as the background onto which a late nineteenth-century audience can project what is missing from the mimesis.

Sex is, therefore, not "depicted" in *Miss Julie* the way the death of the princess and her father are depicted in *Medea*; the former requires of the audience an entirely different sort of image-making, one much less verbally structured on the part of the playwright. Strindberg first installs a gap in the action of the play and then fills that gap not with retrospective narrative but by giving spectators an idea of an actual physical location—"there behind that closed door"—where the missing action may be imagined to occur. Having first been made to believe in an extension of mimetic space just beyond the edge of the stage, spectators then find it easy to imagine that space as substantial enough to support actions they neither see nor subsequently hear described. One can think of the offstage space in this case as a kind of blank screen onto which individual spectators may project whatever degree of vividness that seems appropriate to their own expectations. ("[T]he beholder," says Ernst Gombrich, "must be left in no doubt about the way to close the gap...he must be given a 'screen,' an empty or ill-defined area onto which he can project the expected image" [Gombrich 208].) The early performance history of Strindberg's play testifies to his initial success in "dramatizing" an event neither seen nor described, but, of course, conventions for representing sex in the theatre are much more relaxed now, and that brings its own set of problems. The last time I taught Strindberg's play, several years ago, not a single student among the more than twenty undergraduates in my class in modern drama understood that midway through the play Jean and Julie engaged in sexual intercourse. Living in a world where campus housing is increasingly assigned without any regard whatsoever to gender, they lacked familiarity with the cultural codes necessary to picture what was going on when a man and a woman who knew each other met in the extrascenic space of Strindberg's drama.

Another methodology by which sex is displaced to the offstage aims specifically to induce a voyeuristic response from the viewer or reader. *Women Beware Women*, a seventeenth-century Jacobean tragedy by Thomas Middleton, provides a good example of how even a theatre where sexual desire cannot naturally be depicted can nevertheless be structured so as to cause spectators to see with their minds' eyes what cannot be shown visibly on stage. One of the women in Middleton's play, Bianca, has been recently married to a man many years her senior. Her beauty and relative innocence (she is barely sixteen) attract the attentions of a lecherous Duke, who plots to have sexual intercourse with her. He conspires with a widow named Livia and a panderer (ironically named Guardino) to bring Bianca and her mother to Livia's house, ostensibly out of neighborly goodwill. But under the pretense of taking a tour of an art collection (surely the invitation by Guardino to Bianca to come with him "to see...rooms and pictures" was as humorous a cliché then as it is now), Bianca is confronted by the Duke, at which point she is either seduced or (as some modern readers argue) brutally raped. The sex act itself is naturally not shown on stage, but it is framed for the audience in several ways. First, the Duke states his intentions to Bianca over the course of a lengthy conversation (2.2.315–85) in which he alternately sweet-talks her ("I feel thy breast shake like a turtle panting / Under a loving hand") and threatens her ("I can command, / Think upon that"). As a result when Bianca and the Duke leave the stage together, it is perfectly clear to the audience what is going to happen as soon as they are out of sight.

Second, Middleton encourages the audience to imagine the Duke and Bianca engaging in intercourse by specifying both a setting and a frame of mind appropriate to erotic acts. A crucial factor in this imaginative construction is a speech by Guardino, who comes on stage shortly after Bianca and the Duke have left, to comment cynically on the immorality of the times and also to give the audience an explicitly erotic visual surface against which to project the characters, who are to be imagined at that moment to be engaging in intercourse somewhere in the "offstage":

> Never were finer snares for women's honesties
> Than are devis'd in these days; no spider's web
> Made of a daintier thread than are now practis'd
> To catch love's flesh-fly by the silver wing:
> Yet to prepare her stomach by degrees
> To Cupid's feast, because I saw 'twas queasy,
> I show'd her naked pictures by the way;
> A bit to stay the appetite. (2.2.395–402)

Finally, and most important in the way it assists the audience in constructing an imaginative proof of the intercourse that "happens" while Bianca and the Duke are off stage, is a running conversation between Bianca's mother and Livia that continues throughout the duration of Bianca's and the Duke's scenic absence. Livia and the mother play chess for most of the scene, including that part involving Bianca's initial conversation with the Duke (which takes place in the "above") as well as the time allotted for the offstage intercourse. The women's conversation consists almost entirely of double entendres where the playing of the game of chess becomes a metonymy for what is first seen and subsequently imagined as the game of love played by the Duke and Bianca. The audience is, of course, enabled to "see" this latter game to the extent that they follow Middleton's verbal instructions:

> *Livia*: Here's a duke
>   Will strike a sure stroke for the game anon;
>   Your pawn cannot come back to relieve itself. (2.2.298–300)

And:

> *Livia*: Has not my duke bestirr'd himself?
> *Mother*: Yes, 'faith, Madam;
>   H'as done me all the mischief in this game. (2.2.413–15)

Clearly here, as was the case with the music and peasants dancing in *Miss Julie*, onstage activity is meant to evoke a sense of offstage simultaneity; as Daileader says, what makes such scenes dramaturgically effective "is this very element of nowness, which (especially when coupled with here-ness) results in the illusion that our participation is possible. Move quickly and you might just catch them at it" (Daileader 36).

In contrast to the voyeuristic structuring of *Miss Julie* or *Women Beware Women*, Shakespeare's *Romeo and Juliet* makes use of a different background against which to project an imaginative picture of sexual love. Midway through the play (Act 3, scene 5), the young lovers appear aloft, at Juliet's bedroom window, on the morning after their wedding night. Thus the scene begins just at the point where sex ends, so to speak, and the imaginations of the audience must be directed, therefore, to something that has already taken place:

> *Juliet*: Wilt thou be gone? It is not yet near day.
>   It was the nightingale, and not the lark,
>   That pierced the fearful hollow of thine ear.
>   Nightly she sings on yond pomegranate tree.
>   Believe me, love, it was the nightingale.
> *Romeo*: It was the lark, the herald of the morn;
>   No nightingale. Look, love, what envious streaks
>   Do lace the severing clouds in yonder east.
>   Night's candles are burnt out, and jocund day
>   Stands tiptoe on the misty mountain tops.
>   I must be gone and live, or stay and die.
> *Juliet*: Yond light is not daylight; I know it, I.
>   It is some meteor that the sun exhales
>   To be to thee this night a torchbearer
>   And light thee on thy way to Mantua.
>   Therefore stay yet; thou need'st not to be gone.
> *Romeo*: Let me be ta'en, let me be put to death.
>   I am content, so thou wilt have it so.
>   I'll say yon grey is not the morning's eye,
>   'Tis but the pale reflex of Cynthia's brow;
>   Nor that is not the lark whose notes do beat

The vaulty heaven so high above our heads.
I have more care to stay than will to go.
Come, death, and welcome! Juliet wills it so.
How is't, my soul? Let's talk; it is not day. (3.5.1–25)

In contrast to Middleton's double entendres or the symbolism of
Strindberg's music and dancing, the foregoing scene is more remarkable
for what it does not say about the young lovers' night of passion than for
what it does; as Daileader notes, what strikes one about this conversation "is its failure (if one can apply the word to such exquisite poetry) to
address explicitly what has just taken place" (Daileader 42). One could
say that for Shakespeare, the value of mimesis through diegesis is here
not limited to the indirect staging of offstage (or unstageable) events.
Far better than any direct reference to Romeo's and Juliet's sexuality,
this scene bypasses any narrative of sex to describe the young lovers'
submerged sadness for a situation they—and we—could not possibly have known they could feel so sad about. Instead of hearing about
their lovemaking, instead of being asked to imagine it, our attention
is drawn to the sound of birds, burnt-out candles, morning mists, and
the coming of light to the sky. The scene does not so much substitute
for what takes place in the offstage as to stimulate our imagination to
lament love's passing, even before we knew it was gone.

With Arthur Schnitzler's *Reigen* (1896; published sometimes in trans-
lations as *La Ronde* or *Hands Around*), the "screen" for picturing sexual
intercourse is somewhat different from the one implicit in *Miss Julie* or
in either of the two Renaissance plays. Schnitzler's drama is composed
of ten independent scenes; each of these "dialogues," as Schnitzler called
them, includes an episode of sexual intercourse. The first scene involves
a soldier and a prostitute; the second, the soldier and a housemaid; the
third, the housemaid and a young man; and so on until the final scene
between a count and the prostitute of the first scene, bringing the play
and its characters full circle. There are representative types from nearly
every rung of fin-de-siècle Viennese society—interior scenes feature a
middle-class couple, the "sweet girl" (a familiar type popularized by
Schnitzler as well as by his contemporaries), a poet, and an actress.
The result is a funny but profoundly discouraging satire. None of the
characters is identified by name; they are merely defined (in the manner of the commedia dell'arte) by occupation or class, and though in
the heat of the moment they all profess love to whoever happens to be
their partner, none of them, from the lowest member of society through

those who belong to its highest strata, is motivated in his or her pursuit of pleasure by anything but the most base of motives, whether greed, or selfishness, or lust. In fact, the higher one climbs on the social register, the more hypocrisy one encounters. Only the prostitute, alone among the entire cast of characters, makes no pretense that "love" in this day and age is anything more than sexual intercourse.

Any one of these individual portraits is comic, crudely realistic. Yet in the aggregate their effect is distinctly melancholy. So perfectly choreographed are language and event that as the play progresses the characters' actions seem less and less humorous and more and more desperate, even nihilistic. The men and women are little more than figures in a puppet play, moved by unseen strings, and the patterns they repeatedly and unconsciously trace give the play a distinctly artificial—if clearly morbid—tone, somewhat like a *Totentanz*. The gaps that occur midway through each scene—marked in Schnitzler's text by rows of dashes—do not create spaces for realistic image-making; to decipher them as spaces for sexual intercourse is a distinctly intellectual rather than affective experience. It is a rendering of dramatic action that attends less to the vivacity of the offstage and more to its potential as aesthetic experience.

Reigen was written during the winter of 1896–97 (making its composition almost simultaneous with the premier in December 1896 of Alfred Jarry's scandalous *Ubu Roi*). But Schnitzler did not publish his work until 1900 when he had printed at his own expense a limited edition of 200 copies to give to friends. Even without publicity, however, *Reigen* soon achieved notoriety. In response to its growing popularity, a commercial run of 40,000 copies appeared in 1903, and "dialogues" four through six were performed that year by a student theatre in Munich. The performance resulted in government sanctions against the theatre, and the following year the book was banned in Germany. In the somewhat more liberal climate after World War I, *Reigen* still provoked controversy: the play was shut down after its premier performance in February 1921 in Vienna, nominally because of the public disturbances it provoked, and the tumult that broke out over a performance in Berlin in June of that year resulted in a famous court case. After a trial lasting six days, managers, directors, and actors were found not guilty of causing a public nuisance and committing obscene acts. During the trial interest focused especially on what was neither said nor seen, the elisions during which time intercourse is understood to take place. In the Berlin performance, these episodes were marked by a lowering of the curtain and some few seconds of music, apparently a *valse triste*. The prosecution alleged that these musical interludes were intended to imitate the rhythm of sexual intercourse.

It was this allegation that prompted Alfred Kerr, the distinguished drama critic, to shout in open court that in that case it was the musicians and not the actors who should be in the dock. Kerr's defense of Schnitzler's text seems like an enlightened display of common sense, but the particular claim alleged by the prosecution should not be confused with the much less radical claim that the music heard during the interludes is indeed intended to stand in for erotic activity. It can scarcely be otherwise. Even if the music has no specific mimetic content, the gap that opens up in the action during each scene is without doubt an invitation to the audience to imagine *something*, and it is inevitable under the circumstances that an audience will take the music to be the inspiration for its daydreams. In and of itself, the music they hear may or may not have anything to do with human sexuality, but in the context of the performance, *any* piece of music, regardless of its kind or content, will provide an audience with all the instructions it needs to perform an act of imagining. This kind of radical substitution of some simple material entity for a grander but absent reality happens all the time in theatrical performances—more so, of course, in the case of a fully semiotized stage such as in Chinese theatre, where, as one scholar puts it, "flags represent whole armies, where a circular movement around the stage indicates a long journey, where a single table and chair can be anything from an inn to a royal palace to a courtroom to...whatever" (Schechner 11).

But this kind of artistic shorthand commonly takes place even on more realistic stages. (Ben Jonson once complained about playwrights who pretended that an actor waving a wooden sword could reasonably stand in for the combatants in a major land battle.) During a performance of *Reigen*, as soon as a pair of characters temporarily slips from view, what they are imagined to be doing inevitably tends to be derived from the particular qualities of whatever piece of music happens to be offered as a proxy, so to speak, for the action we cannot see. If, for example, during one of the hidden sexual episodes, a trumpet blares loudly and then humorously falters, as occurs during one scene in Max Ophül's film version of Schnitzler's play (*La Ronde*, 1950), one is then quite likely to conclude that the diminishing sounds are meant as a frivolous caricature of a detumescing penis. Kerr's objections are misleading, therefore: he wrongly assumes that seeing in the theatre is dependent only on mimesis, but the inner logic of the substitution of music for action or narrative is such that it demands an audience associate music with sex—even to the extent of such specific associations as were alleged by the prosecution. The connection between the two is clear and open, requiring no prompting whatsoever: knowing what to look for, so to speak, when the curtain falls, one knows exactly what to "see" in response to the music.

Nor does the music have to be explicitly sensual to function as a suitably imaginative prompt for the audience. Arguments such as that advanced by Kerr seriously underestimate the capacity of the mind to fit whatever substitutes are at hand to the task of imagining. It often happens that the greater the distance between the stimulus and the imagined event, the more strikingly that absent reality can be made to seem present. William Wycherly made china synonymous with sex in a famous scene of *The Country Wife*; nor has anyone ever seriously believed that Jan Dutton was talking about the sport of table tennis when in *You Can't Cheat an Honest Man* the actress turned to W. C. Fields and asked, "How is your ping pong?" It is naïve to deny that these scenes are expressly ribald because in reality sex has nothing to do with china or ping pong; in fact, it is entirely appropriate that the real subject of both conversations, sex, be symbolized by dissimilar entities because it is precisely the incongruity of sex and these things—china and ping pong!—that makes them stimulating to the imagination in the first place. Making these absurd connections offers spectators a type of pleasure like that of deciphering puzzles or contemplating the wild analogies in metaphysical poetry.

Schnitzler's method of dramatizing offstage events lacks the intricate, step-by-step verbal constructions that typify the narratives of classical messenger scenes. Although the playing of the music invites spectators to imagine an activity, the object of that imaginative creation is qualitatively different from what is pictured in the mind as a result of listening to a messenger's report or as a consequence of hearing sounds or conversations emanate from an apparently real but invisible space. One does not imagine the characters in *Reigen* having sex with quite the same completeness one imagines the destruction of the princess in *Medea*. In *Reigen*, with only the music and the lowered curtain to guide it, the mind's creations are much less vivid because they take place in a space only vaguely conceived. The situation would be different, of course, if the events that one supposes to be occurring offstage were defined by more specific and expressly material attributes—say, if the characters were to discuss their encounter in detail after the curtain had come back up, or if we were to hear them making noises offstage, as we hear, for example, the sounds of Agamemnon's murder. (Something like this happened during at least one performance of the play: in Berlin in 1921, the actors seem to have stayed silent during the interludes, but Virginia Woolf, after having witnessed a private showing of *Reigen*, is said to have complained that the performance was "spoiled" because the actors' offstage groans made the audience feel as if a real copulation was going on in the room.)

Schnitzler's play offers a formula for staging the erotic that proves remarkably successful, even when the social climate is such as to allow playwrights room to develop a much more sexually explicit dramaturgy. David Hare resurrects Schnitzler's formula for the erotic almost without alteration in his popular contemporary version of *Reigen*, *The Blue Room* (1998). Hare claims that *The Blue Room* is "freely adapted" from Schnitzler's play, which might seem to imply that he intended to produce a bolder, contemporary treatment of Schnitzler's work. The setting of Hare's play is generalized: it takes place, he writes, "in one of the great cities of the world, in present day." Accordingly, his characters and their conversations reflect contemporary urban life: the soldier is replaced by a cab driver, the maid by an au pair, the poet by a playwright, and so on. But the ingredients for representing sex are almost exactly as Schnitzler invented them a hundred years earlier. Each scene centers on the "absent presence" of sexual intercourse, and each time that event is "dramatized" by a brief interlude of darkness and music. Much of the apparent sexual frankness of the contemporary play comes from its emphatic obscenities:

> *Cab driver*: I don't get it. What am I meant to say? What do you want
> me to say?
> I'm offering you a fucking lift. (Hare 11)
> *Student*: I'm fucking a married woman. (Hare 30)
> *Model*: He was a cunt. He was a cunt, or why else would he have
> left me. (Hare 46)
> *Playwright*: I worship you. I fucking worship you, my child. (Hare 58)

Another difference between Hare's dramaturgy and Schnitzler's is that Hare adds to the interlude a screen projection on which is printed the time one is to imagine having lapsed while the characters engage in sex. Depending on which particular couple is to be imagined offstage, one reads, variously, "three minutes," "nine minutes," "thirty two minutes," and so on, all the way up to a prodigious "two hours and twenty-eight minutes." Yet for all its panache, *The Blue Room* is in some ways tame in comparison with Schnitzler's play. Schnitzler gives his audience absolute freedom to daydream, but Hare, but directing spectators' imaginings of what occurs offstage to the slender resources of a clock, deprives the play of those qualities that might enable an audience to picture the lovers in their full weight and solidity in favor of a metonymic abstraction. The screen projections add humor and irony to the episodes—during a performance, it becomes something of a sport for the temporarily unsighted spectators to learn which couple performs longest. Thus by the end of the play the various pairs have all been placed somewhere on a scale that ranges between the comically inept ("0 minutes") to "Two hours twenty-eight minutes," an achievement that would be heroic were it not apparently drug-enhanced. The image of the stopwatch, by its very frivolity, leads one to conclude that the offstage activities are mechanistic rather than genuinely erotic.

That Hare assigns the most protracted lovemaking to the Politician and the Model may well be part of his satire on contemporary culture. In much the same way, Schnitzler depicts the relationship between the husband and wife as the most false. The allure of the unseen in *The Blue Room* differs in no essential way from that of its century-old antecedent. Indeed Hare's play provides ample demonstration of the way the phenomenon of the erotic in the theatre can be produced effectively by a structure of deliberate concealment. By denying sight of the object, playwrights may actually cause audiences to see it. The principle in this case is the commonplace that eroticism is not simply a matter of the frank display of the human body, whether breasts or limbs or genitalia, but precisely a matter of their strategic concealment. "The erotic photograph," says Roland Barthes, "does not make the sexual organs into a sexual object; it may very well not show them at all; it takes the spectator outside its frame…as if the image launched desire beyond what it permits us to see" (Barthes 59).

CHAPTER 2
AGAINST MIMESIS

"But if the poet should conceal himself nowhere, then his entire poetizing and narration would have been accomplished without imitation. . . . If Homer, after telling us that Chryses came with the ransom of his daughter and as a suppliant of the Achaeans but chiefly of the kings, had gone on speaking not as if made or being Chryses but still as Homer, you are aware that it would not be imitation but narration, pure and simple. It would have been somewhat in this wise. I will state it without metre for I am not a poet: the priest came and prayed that to them the gods should grant to take Troy and come safely home, but that they should accept the ransom and release his daughter, out of reverence for the god; and when he had thus spoken the others were of reverent mind and approved, but Agamemnon was angry and bade him depart and not come again lest the sceptre and the fillets of the god should not avail him. And ere his daughter should be released, he said, she would grow old in Argos with himself, and he ordered him to be off and not vex him if he wished to get home safe. And the old man on hearing this was frightened and departed in silence, and having gone apart from the camp he prayed at length to Apollo, invoking the appellations of the god, and reminding him of and asking requital for any of his gifts that had found favour whether in the building of temples or the sacrifice of victims. In return for these things he prayed that the Achaeans should suffer for his tears by the god's shafts. It is in this way, my dear fellow," I said, "that without imitation simple narration results." "I understand," he said.

VII. "Understand then," said I, "that the opposite of this arises when one removes the words of the poet between and leaves the alternation of speeches." "This too I understand," he said, "it is what happens in tragedy."

(The Republic, Book 3, vi–vii)

The foregoing passage from *Book 3* of *The Republic* is often overlooked in favor of Plato's more famous criticism of poetry in *Book 10*, where he condemns tragic poets not only because their imitations are three degrees removed from God and truth, but also, he says, because "poetry feeds and waters the passions instead of drying them up; she lets them rule, although they ought to be controlled." In Plato's view, to be sure, it is not just poetry that ought to be banished from the ideal state; all mimetic art is suspect for the reason that people who are exposed to it are deceived by the alluring qualities of the represented figures or images. Yet Plato aimed his criticism mainly against those poets who wrote for the stage, and it is here that his influence has been felt most keenly in subsequent discussions about the proper place of the arts in human societies. His insistence that spectators to theatrical performances are somehow unable to shake off habits of behavior they picked up from watching dramatic performances, whether tragic or comic, has been carried forward virtually unchanged for well over two thousand years of antitheatrical polemic. It forms the basis of every major subsequent antitheatrical discourse, being repeated (usually without acknowledgment of its ancient roots) in the seventeenth-century Puritans' or Jansenists' attacks on theatre or in the contemporary debate about the influence of hip-hop music. At its core are basic realities of human nature, possibly of the primate brain itself: we seem to be born hardwired for empathy, so to speak, so that we are inevitably moved in some ways to respond to others' experiences as if they were our own (Fountain, *The New York Times*, August 24, 2004, D1). As a result, we are easily aroused by most kinds of dramatic art to the highest levels of pity, sympathy, and fear, and we can readily be moved to feel rage, or jealousy, or sorrow, or sexual attraction, by the behavior of actors on stage; moreover, whenever this happens, we seem not to notice, or not to care, that our emotions are being spent on simulacra. So we tend to treat the figures we see on stage (or on screen) just as if they were real people. We are happy in their company, or irritated by their shortcomings; especially if we like them, we are buoyed by their successes and saddened by their failings, counting both in some ways as our own; we fall in love with them, we idolize them, obsess over them, and in general extend to them every quality or emotion commonly associated with real human beings, including the possibility of our becoming permanently changed for better or worse by affective involvement with them.

The universality of mimetic performance, the ease and thoroughness with which we instinctively respond to it, immediately suggests that an important cultural functioning for the imitative arts is being expressed with such behavior. Indeed recent neurophysiological research data, according to Ellen Dissanayake, provide evidence for just this kind of mimetic responsiveness to all the arts. "[T]he work of art," says Dissanayake, "writes itself on the perceiver's body: electrochemically in the signaling patterns of activity that comprise the brain's cortical maps, which may in turn have concomitant physiological and kinesthetic effects. The sensation (in bones and muscles, in the being) that the empathists wished to explain—of union or communion between viewer and object, listener and musical work, reader and poem—is real, not illusory or only metaphorical" (Dissanayake 185). That the mimetic arts might play an informative or even decisive role in human emotional development is a position, paradoxically, that the critics of theatre take up more vigorously than its proponents. The reason that theatre in particular has generated such fierce opposition, historically speaking, is precisely that its critics believe that it is better than the other modes of literary art at the mimesis it performs and that it stimulates audiences to perform as well. Thus a writer like Thomas Dekker, in documenting the case against Elizabethan theatre in his *Gull's Hornbook*, noted the ease with which the behavior of the audience could be made to match that of the players: "men come not to study at a Play-house," Dekker wrote, "but love such expressions and passages, which with ease insinuate themselves into their capacities" (Nagler 131). Yet with a few notable exceptions (Barish, *The Antitheatrical Prejudice*; Puchner, *Stage Fright*), centuries of learned commentary on the theatre has tended to belittle sentiments such as Dekker's, along with the huge accumulation of similarly hostile accusations against theatrical performances, as uneducated or crude. Even when apparently real empathic responses such as the ones Dekker observes are documented (as they are, for example, in this and other Elizabethan commentaries on actors and acting), such behaviors are often said by modern scholars to refer to "popular" reactions to dramatic performance, these crude responses to be distinguished from more restrained, legitimate ones (Nagler 113–38). Likewise, when contemporary public figures speak out against the effects of mimetic performances, their "copycat" theories are dismissed as lacking merit. When, in 1992, Vice President Dan Quayle criticized the then popular television series "Murphy Brown," where Candice Bergen portrayed a single woman who decided to have a baby without first getting married, few took his arguments seriously. For asserting that thousands of young women would learn irresponsible behavior from Bergen's portraiture, Quayle was widely mocked as a national fool.

But what if we encounter in more cautious thinkers many of the same suppositions with regard to the likelihood that publicly performed mimetic acts are capable of bringing about long-term behavioral changes? Would there then be reason to take theatrical performances and their critics a little more seriously? In particular the copycat

argument, long dismissed in academic discussions of theatre as unre-
fined or unintellectual, seems a little more plausible when it is advanced
by people with unimpeachable experience and accomplishments in the
performing arts. Ironically, a number of twentieth-century playwrights
and theatre artists provide a good deal of supporting testimony for
exactly the kind of affective potency that theatre's most narrow-minded
critics have long ascribed to it. Even though the majority of theorists
of literature since Aristotle have tended to assume that the mimetic
responsiveness of audiences was on the whole beneficial, the great iconoclastic thinkers on the subject of the truthfulness or the utility of
mimetic representations, from Plato to Jacques Derrida, have always
taken their bearings from a theoretical point of view that assumes that
a proper response to theatre art requires some way to balance mimesis
with something like what the philosopher Edmund Husserl called acts
of "relational contemplation." In the early decades of the twentieth century, for example, the British theatre artist Edward Gordon Craig vigorously and repeatedly defended the office of the Censor, arguing that
it was necessary to protect innocent theatergoers from seeing "jealousy,
ambition, hatred, meanness, envy, murder, torture, seduction, and riot"
(Craig 148). Likewise Brecht thought that the theatre had great power
to shape both politics and character, and he attributed that power to its
empathic appeal in much the same way that Plato did:

> Neither the socialist thinker nor the parson in his pulpit would deny
> that our morals are affected by it [the theatre]. It matters how love, mar-
> riage, work and death are treated on the stage, what kind of ideals are set
> up and propagated for lovers, for men struggling for their existence and
> so on. In this exceedingly serious sphere the stage is virtually function-
> ing as a fashion show, parading not only the latest dresses but the latest
> ways of behaving. (Brecht, *Brecht on Theatre* 151)

There is no evidence that Brecht knew of Craig's writings on theatre,
the majority of which were published when Brecht was still a school-
boy. Yet Brecht and Craig share a similarly critical view of the way
conventional acting styles influence the hearts and minds of audiences;
"[Craig's] argument that the actor should have his 'brain command-
ing his nature,'" writes J. Michael Walton, "is as clear an exposition
of the Brechtian principle as one could hope to find, and pre-dates
Brecht by twenty years" (Walton 73). I want to suggest that Brecht's
and Craig's general views of the psychological effects of mimesis are
expressly Platonic, and I want to suggest further that their ideas for

theatre—even though there is no question of direct influence—are part of a larger historical and specifically modern preference for diegetic conversions or displacements in theatrical performance (Gruber, "Building an Audience" 72).

The argument against mimesis that Plato advances in Book 3 of *The Republic* is potentially the more important of his two sustained commentaries on poetry. In it Plato not only attempts to discredit the imitative arts because of their remoteness from God and truth but also includes suggestions on how to modify the undesirable effects of mimetic performance by imbuing it with some of the formal properties and characteristic effects of narrative. The distinction he makes turns on a difference between two different kinds of poetic performance, based, respectively, on mimesis and diegesis. Of these two analytic terms, mimesis is by far the more familiar, owing in large part to the fact that in his attempt to respond to his teacher's condemnation of poetry, Aristotle had installed imitation as the formal method of art. But the general concept of poetry as imitation has even earlier origins in Greek thinking about the arts, and it would be wrong to view Plato's understanding of the psychology of mimesis as unusually biased or idiosyncratic. As early as the fifth century B.C., for example, in what is one of the earliest instances in which poetic creation is defined as an imitative act, a character in Aristophanes' comedy *Thesmophoriazusae* remarks that the successful tragic poet is one who becomes one with his characters, whether male or female; whatever qualities he doesn't possess by nature, he contrives to acquire by imitation (*ha d'ou kektêmetha, mimêsis êdê tauta sunthêreuetai*; Aristophanes 144).

As for the term "diegesis" (literally, a narration or narrative account), it is clear that Plato introduces it (in the passage quoted in the epigraph to this chapter) as a way to distinguish two different methods by which poets might represent reality. The problem Plato here finds with poetry and poetic performances (as distinct from his treatment of them in Book 10) is less philosophical than formal. When in their performances the poets liken their speech and gestures after those of their characters, says Plato, so that in some sense they seem to become identical with them, they enact a "diegesis through mimesis"—a kind of telling through showing—in which the retrospective enactment of reality is turned into something hard to distinguish from reality itself. A more desirable kind of performance, however, would be simply diegesis, or telling—a "diegesis without mimesis." In this latter case the poet's voice remains distinct from the speech of the characters, so that the actual telling of events is never fully subsumed by the showing; the poet's narrative tags separate events from the cognitive reporting of them (Gruber, "Non-Aristotelian Theater," 208).

Plato's thinking in this remarkable piece of literary criticism is far from thick-headed or naive. For one thing, he is not simply bigoted against theatre, as his general attitude is commonly misunderstood to be. It would be more accurate to describe Plato's opposition toward mimesis as directed specifically (to borrow Theodor Adorno's words in his essay on "committed" art) against "a work of art that is content to be a fetish, an idle pastime for those who would like to sleep through the deluge that threatens them, in an apoliticism that is in fact deeply political" (Adorno 177). In some ways (as is suggested by its resemblance to the theories of Brecht and Adorno), Plato's criticism of mimetic performance anticipates a number of twentieth-century dramaturgical styles that develop in opposition to strictly representational acting traditions and to conventional models for mimesis. His ideas provide a theory of drama expressly suited to the modern period in which the pioneering works of theatre are political rather than hermetic, popular rather than elitist, and synthetic rather than formal. His writing on narrative interventions especially provide a useful conceptual frame for the work of twentieth-century playwrights whose interests and dramaturgies are as different as those of Yeats and Brecht (Gruber, "Non-Aristotelian Theater" 212). In simplest terms, like Plato, they want a theatre that modifies "showing" with conspicuous elements of "telling."

In the above passage, for example, Plato explains how the formal markers of narrative—the "he says," so to speak—may be included in the poet's performance in order to establish what amounts to a series of checks and balances on purely mimetic representation. Socrates differentiates two kinds of poetic performance, one in which the poet relies mainly on indirect discourse and another, a contrasting style more clearly identified with drama, in which one omits the intermediating words of the poet and leaves only the actual words of the characters. Within the former style, according to Plato, the poet's role as a narrator confines the imitated events within a coherent structure of representing; his stance merely as a reporter or recorder of absent events is consistent and clear. Plato strongly prefers those performances that include formal elements that somehow frame, qualify, interrupt, or even contradict the effects of mimesis; this is the kind of performance style he calls diegesis, and the common discursive markers of that style—what Plato calls the "he says"—establish a categorical separation between original speech and its subsequent representation. Socrates contrasts this manner of poetic performance with performers who omit the "he saids" and speak the words as if they were their own; such uniformly dramatic performances inevitably blur the distinction between the illusions of mimetic enactment and pure, original event.

Plato sees conventional narrative markers as small but critical modifications to mimetic performance. They establish within the performer's imitation a steady rhythm of interruption and, more important, they show a conscious commitment to some cognitive frame beyond the imitation. To the extent that the poet frames showing with telling, Plato seems to be saying that spectators necessarily remain aware of the constructed nature of the performance. One could say that whereas in mimetic performance the events are transmitted from within a communicative system that depends on the illusion of "here and now," the kind of performance that Plato is advocating carries with it the marks of the poet's attitude toward his utterances. Unlike a purely mimetic performance, such a performance has inherent in it the quality of speech acts that linguists call "modality" (or "modalization"); within such a context, in other words, the poet produces his or her discourse while simultaneously acknowledging the "saying" of the text (Pavis 216). Under these conditions, according to Plato, spectators are less likely to succumb temporarily to the performance's illusory reality.

The privileging of narrative over enactment can take other forms as well and serve alternate purposes. Socrates' description of his unprofessional manner—"I will state it without metre for I am not a poet"—is also a crucial part of the diegetic style that Plato intends. Paradoxically, for Socrates, *not* being "poetic" is actually a desirable quality for a poet's performance for much the same reason that narrated imitations are better than actual dramatizations. As the narrator's own words provide a kind of check on mimesis by regularly interrupting the speeches of the characters, so a measure of awkwardness on the part of the poet makes it all the more unlikely that the performance will be mistaken for the real thing. From the point of view of one who advocates a high degree of cognitive awareness in responding to mimetic art (as Plato surely does), one could say that a certain degree of ineptness has value for the conspicuous disharmony it brings to artful representations. Brecht said as much, early in his career in an essay on "unprofessional" acting, that he saw material poverty as a virtue because a theatre with limited financial resources could not afford to pretend that its theatrical productions were anything other than theatrical productions. But a theatre with a lot of money to spend on sets and set designers could much more readily transform diegesis into mimesis: "[a] little more money," Brecht writes, "and the room shown on the stage would be a room" (Brecht, *Brecht on Theatre* 149). Brecht's impoverished set functions in much the same way as Socrates' "unpoetic" performance; in both cases the material shortcomings undercut the imitation by making its status *as* imitation readily apparent.

A number of other twentieth-century playwrights also seem to view with skepticism a theatre that is purely mimetic, and they tend to favor stylized or obviously artificial performance styles for some of the same reasons that Plato recommends interrupting enactment with narrative tags. Martin Puchner has shown that a specifically Platonic "antitheatricalism" forms an important strain of modernism in the theatre, a theatre—as Plato might have put it—"without theatricality" (Puchner 7). The most well-known example of a twentieth-century playwright who was overtly hostile to conventional mimesis would, of course, be Brecht, who spent his career trying to separate the theatre from its economic and political underpinnings and whose ideal for the "epic" style (as Walter Benjamin long ago noted) held up a peculiarly retrograde, "Platonic" model for theatre (Benjamin 85–103). But Brecht was merely the most outspoken among a large number of playwrights who were critical of a theatre built exclusively on mimesis and empathy.

There is no case to be made for Plato's "influence" on modern drama, since his writings on dramatic performance have been for the most part ignored by playwrights and commentators, and they remain, until fairly recently at least, largely untheorized by scholars of drama who have typically disregarded them when they have not disparaged them outright. Yet Plato's idea for a poetic performance that mixes diegesis with mimesis anticipates a large variety of twentieth-century dramaturgical styles (Gruber, "Non-Aristotelian Theatre" 212). This correspondence of modern aesthetics with ancient mimetic theory can be identified early in the twentieth century in the experimental theatres of high modernists such as Gordon Craig or W. B. Yeats. Platonic models for a diegetic theatre can be seen as well in avant-garde pieces composed by Gertrude Stein, or, still later, in some of the haunting narratives and dramaticules of Samuel Beckett, which typically inhabit a literary domain somewhere between "story" and "play." And some form of a Platonic, "mediatized" stage exists in much postmodernist theatre, where diegetic theatrical forms seem to have become almost a default position in the work of artists such as Peter Handke, Richard Foreman, Suzan Lorie Parks, or Robert Wilson (Puchner 173–74).

In the work of some of these playwrights the diegesis derives as much from the mise-en-scène as from the text; the action in Parks' *Topdog/Underdog* (2001), for example, includes some of what Parks calls her "slightly unconventional theatrical elements" (Parks 1708). These include numerous "rests"—distinct pauses or "beats" of silence that suspend the action—and "spells," which, according to Parks, are "elongated and heightened rests" that are marked in the text by the repetition of the characters' names (Parks 1708), as in the following passage:

> *Booth*: You a limp dick jealous whiteface motherfucker whose wife dumped him cause he couldnt get it up and she told me so. Came crawling to me cause she needed a man.
> (*Rest*)
> I gave it to Grace good tonight. So goodnight.
> *Lincoln* (Rest): Goodnight.
> Lincoln
> Booth
> Lincoln
> Booth
> Lincoln
> Booth[.] (Parks 1718)

Silences such as these on stage are too long and too frequent to be purely mimetic; they are difficult if not impossible for an actor to fill in with a motivational subtext, as, for example, the scripted silences in the plays of Chekhov are typically covered over; they have, instead, as Parks notes, "sort of an architectural look" (Parks 1708). In this case the silences, rather than calling upon the actors to invent a subtext to motivate them, tend instead to redirect spectators' attention to the structural matrix of theatrical performance, with much the same effect as the supplemental narrative markers recommended by Socrates. In both cases the intrusions "cut up" the mimesis into somewhat independent elements, rendering the performance as a sequence of independent dramatizations or "quotations."

Brecht, like Plato, expressed a desire for a theatre that rigorously imposed diegesis on mimesis. His methodology is taken up (either consciously or intuitively) by an entire generation of later playwrights such as Howard Brenton, Ntozake Shange, or David Henry Hwang; all are broadly political playwrights whose theatres offset mimesis with what might be called narrative interventions or screens. What is implicit in this overview of twentieth-century theatre, however, is not so much a conscious or even a consistent "antitheatricalism" (at least as that term is used in its conventional sense); rather, there extends through much of the theatre of this period a more diffuse Platonism, a tendency to shift mimetic performance in the direction of storytelling or to temper scenic enactments with some of the qualities of narrative.

With a few modern artists there was sometimes an element of old-fashioned Puritanism in the mix. Craig, for example, was keenly aware of the affective potency of actors and the plastic responsiveness of spectators to conventional mimetic performance, and in order to minimize the tendencies of spectators to feel the range of actors' emotions as their own, he imagined a fundamentally sanitized theatre where aesthetic consciousness largely replaced simple empathic involvement. Craig often displayed an apprehensiveness about dramatic performance that seems as hostile to theatre as any zealot's rant: "An audience," he wrote once, "witnessing 'Othello' will be led to suppose that jealousy is an attribute of the noble-minded and the strong, and that a jealous man is privileged to wind all things up to his own advantage by committing murder and afterwards a glorified suicide" (Craig, *The Mask* 2: 31).

Yeats, on the other hand, was not nearly so overtly suspicious of mimesis as was Craig. But the majority of Yeats' works for the stage nevertheless show a similar bias against conventional forms of scenic enactment; his opposition to the then dominant realist or naturalist theatrical styles can best be understood in terms of a marked preference for the broader, more cognitive responses to theatre advocated by Plato. Indeed, during the early years of the twentieth century Yeats and Craig corresponded enthusiastically with respect to the ideal, abstract, mediated experience of theatrical performances that each seemed independently to be searching for. Yeats had written in 1904, for example, that "there should be something in their [the actors'] movements decorative and rhythmical as if they were painting on a frieze" (Yeats, *Explorations* 176–77), and in *Four Plays for Dancers* (1916), the action is framed both verbally and visually by extrascenic figures (identified by Yeats as "musicians") who chant narrated descriptions of the characters, the plot, and the imaginary setting for the action. Likewise Craig attempted to impose diegesis on mimesis, in order to make theatrical representation not just "showing" but more like "the showing of showing." In his adaptation of Handel's pastoral opera *Acis and Galatea* (1902; Handel in turn had created his work for the stage from Ovid's tale), Craig experimented with a stage décor that blended actors and environment into a single aesthetic continuum. For that production, Craig robed both set and actors in flowing ribbons so as to blend visually the human figures with their physical environment; the effect was visually stunning. Max Beerbohm, for example, wrote in his review of Craig's production (*Saturday Review*, April 5, 1902) of "the fluttering grace of those many-ribanded costumes... [and] the cunning distribution and commingling of figures and colours" (Burden 450).

Craig's *Acis and Galatea* is usually considered to be an attempt to install a formal aesthetic unity on stage performance. In particular Craig intended to minimize the disruptive presence of a living body on an otherwise wholly artificial environment (Olf 488–94). In this endeavor, Craig (like his contemporaries in theatre Adolphe Appia, Vsevelod Meyerhold, and Oskar Schlemmer, among others) was attempting to purge the stage of anything but its own fundamental and unique artistic elements, much as modernist painters attempted to purge their art of anything but its own autonomous formal elements or to develop on their canvases a tension between representation and conspicuous patterning or design. But Craig's extreme formalism can be seen alternatively as a replacement of the mimetic stage with a rigorously artificial design that because of its dominant aesthetic elements tends to minimize the empathic element of the actors' performance, in particular the sad tale of Galatea's love for Acis. In place of a mimesis where the audience responds, to borrow José Ortega y Gasset's words, to "nothing but the moving fate of John and Mary or Tristan and Isolde" (Ortega 10), Craig substitutes the artificial structure of the presentation itself.

These and numerous other stylized, self-conscious uses of diegesis or diegetic techniques can be sharply distinguished from the more strictly mimetic narratives of the classical messenger. The chanting of Yeats' musicians or the extra-dramatic "speaking against the music" that Brecht prescribed for the singers in his *Threepenny Opera* is not, by and large, an instance of narrative devoted exclusively or even mainly to the representation in words of offstage events, as were the narratives of Euripides' messengers. Nor is their purpose (as was the case with the Euripidean messenger) chiefly affective and/or imaginative. Especially important is their relative disengagement from the action: in many cases, the people who perform these narrative roles do not exist entirely within what Manfred Pfister, in his influential study *The Theory and Analysis of Drama*, calls "the internal communication system" of the play

(Pfister 74). Yeats and Brecht pursue a conspicuously "Platonic" dramaturgy; in their works, the narrative elements are more functionally mediatory, negotiating and controlling the "traffic" between audience and representation in much the same way that the "he saids" of Socrates' "unpoetic" recitation frame the mimesis of Plato's ideal performance. In works such as these, the purpose of narrators on stage is not so much to relate unseen events as to bracket mimesis, distancing or estranging it in order to serve aesthetic or political ends rather than ideational or affective ones. Such "mediatized" or "narrativized" dramaturgies—as distinct from the narrative tellings of more conventionally representational styles of performance—are the subject of this chapter.

*   *   *

In order to understand the ways that narrative interventions have been instrumental in conferring something like a distinctive period style on much twentieth-century drama, it will be useful first to review the historical tradition from which the various forms of diegetic elements in drama emerge. Narrative filters (or frames) of the general kind and purpose envisioned by Plato are widespread throughout drama, appearing so often as to be themselves a kind of familiar feature of the genre. Prologues and epilogues, choruses, asides, the commentary of *raisonneurs*, all represent some kind of intrusive supplement to the mimesis, an accompanying narrative component that audiences understand to be separated in some way from the structure of enactment that in theory ought to contain them. Pfister uses the umbrella term "epic communication structures" to distinguish all those components of plays that resist being enfolded within the fictional world (Pfister 74); they include, he says, "prologues and epilogues that are presented by a figure outside the internal dramatic action" as well as the classical chorus, "in so far as it remains a figure-collective outside the internal dramatic system and comments on the dramatic situations without getting involved in them" (Pfister 74).

Epic (or narrative) elements such as these are found so commonly in plays of all eras that their inclusion within the general tradition of stage representations constitutes something of a paradox. Proscribed by Aristotle in *The Poetics* as fundamentally incompatible with drama (tragedy, wrote Aristotle, is "acted, not narrated"), the presence of narrative elements within dramatic enactment, or (as Plato might put it), the combination of "mimesis with diegesis," seems ubiquitous in theatre history. Narrative incursions are the rule, if anything, rather than the exception. As Pfister writes, "the absence of a mediating narrative function represent[s] an idealised norm from which dramatic texts have frequently deviated over the centuries" (Pfister 69).

But mediating against, or for, what? What motivates a dramatist to want to frame, screen, disrupt, or in any other way qualify purely mimetic enactments? Perhaps the most common reason for a playwright to employ an extrascenic commentator would be for the purpose of advertising the play to a prospective audience or to stipulate certain conditions about its setting or performance. This technique is easiest to see in the case of a prologue, where a speaker (who may or may not speak "in character" and who may or may not subsequently participate in the play itself) comes on stage before the play begins to try to sell it to the audience. The Roman playwright Plautus, who had to compete for spectators to his plays with a number of other popular forms of street entertainment, often begins his comedies with an overview advertising his characters and plots in the manner of a modern Hollywood film trailer. Thus in *Amphitryon*, for example, the god Mercury takes the stage for some one hundred fifty lines to entice the audience with a tale of illicit sex in high places, while in *The Two Menaechmuses*, the speaker of the Prologue explains in detail how the twin brothers came to be separated shortly after birth, giving spectators key information without which the ensuing mimesis would make little sense.

Similar narrative frames can be found in medieval dramas such as *The Castle of Perseverance* or *Everyman*; the diegetic materials for the former play are particularly elaborate and dramaturgically interesting, involving a lengthy and intense debate between two figures identified as the first and second "vexillators" (literally, "flag-bearers"), whose function it is to summarize the theological and moral background for the allegory. After describing in lucid and alarming detail the perilous state of men's souls, the vexillators then announce formally their intentions to replace diegesis with mimesis for the rest of the play:

> Grace if God wil graunte us, of his mikyl myth,
> These parcell[ys] in propyrtes we purpose us to playe
> At _____ on the grene, in ryal aray. (Bevington 803)

Likewise sixteenth- and seventeenth-century playwrights frequently make use of an introductory narrator who gives capsule summaries of the plot or adjusts the attitudes of spectators. Even Shakespeare, whose dramaturgy normally tends more toward the mimetic norms for the genre set down by Aristotle, on several occasions employs extrascenic commentators. *Troilus and Cressida, Romeo and Juliet,* and *Henry VIII* begin with a brief choral overview, and all five acts of *Henry V* open with a narrative commentary spoken by a Chorus, which repeatedly calls on spectators to supply with their imaginations what cannot credibly be represented by the actors on the playing space, as here at the commencement of Act 5:

> Vouchsafe to those that have not read the story
> That I may prompt them; and of such as have,
> I humbly pray them to admit th' excuse
> Of time, of numbers, and due course of things
> Which cannot in their huge and proper life
> Be here presented. (5.1.1–6)

If playwrights use narrative incursions such as these to provide background information efficiently to an audience, or to plead spectators' indulgence for any shortcomings in the mimesis, sometimes they use them to ask for special considerations for themselves or for their work. Classical comic playwrights—in what must have been among Western civilization's earliest instances of the commercialized branding and marketing of a product—frequently used their prologues to advertise the superiority of their plays over those of their contemporaries. At other times the playwrights used these prologues to respond in their own voices to the charges of critics, as when, for example, the speaker of the Prologue to Terence's *Andria* complains that the author "wastes a lot of time writing prologues, / Not to unfold the plot, but to reply / To jealous charges from an older poet" (Terence 10). Much the same desire to respond to contemporary criticism prompted Ben Jonson in the winter of 1609–1610 to write a second prologue to *Epicoene* in which he instructed spectators to "think nothing true" in his story; in denying a connection between literature and life, Jonson was attempting to silence widespread accusations that his comedy intentionally satirized Lady Arabella Stuart.

Narrative components of plays such as these (as well as their formal counterparts, choral interludes) clearly belong outside the spaces and times that are represented mimetically in dramas. Whenever they are used by playwrights, they bring about a kind of distancing or mediating effect on dramatic performance that Plato argued was preferable to unmixed mimesis. Indeed, Brecht was to make a textual equivalent of this kind of narrative incursion into one of the mainstays of his epic dramaturgy, using placards, screen projections, or doggerel verses (as are called for in *Mother Courage* or *Galileo*) to frame the various scenes in the play.

A second variant of diegetic restriction on mimesis involves a character who describes events he or she witnesses as they are occurring just beyond the periphery of the represented space. This "blow by blow" account of offstage events (known formally as teichoscopy) is different conceptually from the narratives of classical messenger speeches, much as a sportscaster's live description of a game differs from a newscaster's retrospective summary. The former gives a real-time account, while the latter reports on events as history. Patrice Pavis, in his *Dictionary of the Theatre*, reviews examples of teichoscopy from various playwrights including Shakespeare (*Julius Caesar*), Goethe (*Götz von Berlichingen*), Brecht (*Galileo*), and Giradoux (*Electra* and *The Trojan War Will Not Take Place*). The technique would be familiar for anyone even minimally versed in classical dramaturgies. For example, in one of the comedies of the medieval playwright Hrotsvitha of Gandersheim (who claimed to have written her plays in imitation of Terence), the wicked Roman emperor Dulcitius attempts to rape a number of sooty kitchen pans that he has been deluded into believing to be three Christian virgins. These three women peek through a crack in a door to watch the emperor's folly, describing the action to spectators who cannot see it: "Look, the fool, the madman base, / he thinks he is enjoying our embrace.... / Into his lap he pulls the utensils, / he embraces the pots and the pans, giving them tender kisses" (Hrotsvitha 216).

Variants of teichoscopy can be found throughout works of the modern theatre. Hauptmann includes in *The Weavers* both a "messenger scene" as well as at least one instance in which a character describes the armed combat then imagined to be occurring simultaneously in the streets outside:

> Mielchen (*Puts her head in through the window laughing*): Grandpa, Grandpa, they're shootin' with guns. A couple of 'em fell down. One of 'em turned 'round in a circle—'round and 'round like a top. One's all floppin' like a sparrow with its head tore off. Oh, and so much blood spurtin' out—! (Hauptmann 155)

Another variation of teichoscopy appears also in Strindberg's *Miss Julie*, when Jean carries on a conversation via a speaking tube with Julie's father, who is imagined to be located in his room offstage. Jean's anxiety is all the more striking because he is so obviously unnerved by the voice of the unseen count. In a scene that underscores, ironically, the accuracy of one of Plato's admonitions against mimetic performance, namely, that it depends for its effect on the unmediated transmission of the words of someone who is not physically present, Jean responds to his master's voice as if he were actually standing before him, much like the dog listening attentively to the phonograph recording in the old RCA logo.

Even more experimental dramatists make use of narrators whose accounts are contemporaneous with the events they describe. In Yeats' early work *On Baile's Strand* (1904), a character identified as "The Fool" relates to a blind man the things he sees taking place in the space outside the door, which is to say out of sight of the audience. At the climax of the drama, Cuchulain, a legendary Irish hero, races offstage in a delusional rage and tries to attack the ocean, which in his madness he takes to be his enemy Conchubar; the scene is not enacted, rather narrated by a character who remains onstage:

> There, he is down! He is up again. He is going out in the deep water. There is a big wave. It has gone over him. I cannot see him now. He has killed kings and giants, but the waves have mastered him, the waves have mastered him! (Yeats, *Eleven Plays* 43)

And in Maria Fornes' *Fefu and Her Friends* (1977)—a zany, surreal drama in which the audience is divided into four groups who take turns moving to four separate spaces where the play's four scenes are being performed simultaneously—the main character, Fefu, takes potshots at her offstage husband, Phillip, who is said to be walking on the terrace outside their house. After she aims the gun through a pair of French doors and shoots, Fefu nonchalantly describes what she sees to her friends: "There he goes. He's up. It's a game we play. I shoot and he falls. . . . It's not too bad. He's just dusting off some stuff. (*She waves to Phillip and starts to go upstairs.*) He's all right. Look" (Fornes, *Fefu* 11).

In the foregoing cases the narrator performs his or her role without seriously undermining the conventions of dramatic mimesis. Even the two modern plays conform to the naturalistic décor developed for the stage during the late nineteenth century; from the point of view of aesthetic technique, the teichoscopies in *On Baile's Strand* and *Fefu and Her Friends* are consistent with the common nineteenth-century scenic practice of placing windows or open doors on the back or side walls of an interior set. Gazing through and beyond these openings in the set and seeing a painted landscape or parts of rooms down a hallway, spectators can readily accept the illusion that the fictional world possesses extensive dimensions similar to the real one. Though teichoscopy is technically a form of diegesis situated within the mimesis, for the most part it has been used, as the previous examples show, for somewhat different purposes and effects than the extended narratives of the messenger in classical tragedy. Like a messenger speech, it eliminates unseemly or problematic acts from view. But at the same time the technique gives spectators a "real time" account of events as well as the chance to exercise their imaginations to create a specific mental image of the extrascenic space as "next door" to the space visible on stage.

But twentieth-century playwrights often employ narrators, whether traditionally retrospective or teichoscopic, in ways that are more boldly innovative and expressly self-conscious than in the previous several instances. When Yeats introduces narrators in some of his later dramas (such as the *Plays for Dancers*), he uses them less to promote verisimilitude or empathy than to provide for aesthetic or emotional estrangement. In such works the narrative interventions are consistent with a counter-theatrical and broadly Platonic project. In the stage directions to the first of his *Plays for Dancers* (*At the Hawk's Well*, 1916), for example, Yeats recommended that the performance take place under normal household lighting. Here his intent was not to have actors and audience share the same physical space and ambient lighting in order to bring them closer together emotionally but to force them into an unaccustomed proximity. This unfamiliar closeness would then underscore the actual psychic distance that exists between performers and their audience: "These masked players seem stranger," Yeats wrote, "when there is no mechanical means of separating them from us" (Yeats, *At the Hawk's Well* 429).

In contrast to teichoscopy, where a narrator substitutes verbal description for direct action or scenic enactment, a second very different and much more characteristically modern technique comes about when such a narrative does not replace enactment but instead *accompanies* it. In this case the speaker does not provide spectators a way to imagine what is understood to be taking place in the offstage; rather his words describe an action that is already apparent and happening in full view. This recursive role for narrative as a component of theatrical performance is practically nonexistent on pre- and early modern stages, but the practice is fairly common in the twentieth century. Yeats employs it, for example, in his symbolist drama *At the Hawk's Well*. The play takes place during the Irish Heroic Age. Like the earlier *On Baile's Strand, At the Hawk's Well* depicts events drawn from stories about the Irish folk-hero Cuchulain, whom Yeats identifies in the dramatis personae simply as "a young man." The plot tells mainly of Cuchulain's quest to find the waters of immortality. The drama begins with an elaborate ceremonial induction: three "musicians," their faces made up to seem as if they are wearing masks, sing an invocation to the audience in which they urge them to use their imaginations in picturing the scene. While they are singing the musicians ceremoniously spread out between themselves a large black cloth ornamented with a gold figuration of a hawk. While this cloth is being opened—this act is Yeats' symbolic representation of the stage curtain being raised—a figure robed in black and represent-ing the Guardian of the Well enters the playing space (which Yeats says can be "any bare space"). She crouches on the ground next to a square blue cloth that represents the waters of the well. As soon as she is posi-tioned, the musicians roll the black and gold cloth back up and retreat to positions on the periphery of the playing space, where they take up a narrative (or choral) role, alternately singing and speaking as they pro-vide the audience more details about the characters, setting, and plot:

> First Musician (*Speaking*) Night falls;
> The mountain-side grows dark;
> The withered leaves of the hazel
> Half choke the dry bed of the well;
> The guardian of the well is sitting
> Upon the old grey stone at its side,
> Worn out from raking its dry bed,
> Worn out from gathering up the leaves.
> (Yeats, *At the Hawk's Well* 429–30)

As the musicians continue to sing, speak, and play on their various instruments (a drum, gong, and zither), an actor wearing the mask and costume of an old man makes his way toward the front of the stage. He moves slowly toward the square of blue cloth that represents the well. As he makes his way across the stage, one of the musicians describes his progress:

> First Musician (*Speaking*) That old man climbs up hither,
> Who has been watching by his well
> These fifty years

> He is all doubled up with age;
> The old thorn-trees are doubled so
> Among the rocks where he is climbing. (Yeats 430)

This and the First Musician's previous speech convey a good deal of background information that is necessary if the audience is to picture what is happening on an otherwise empty stage. The First Musician's narrative is essentially expository; in it he asks the audience to picture a specific natural environment, and he identifies a masked character unknown to the audience and provides him with context—time, place, and motive. Yeats' text also describes the actions the actor who plays the Old Man is to perform: "*The* Old Man *stands for a moment motionless by the side of the stage with bowed head. He lifts his head at the sound of a drumtap. He goes toward the front of the stage moving to the taps of the drum. He crouches and moves his hands as if making a fire. His movements, like those of the other persons of the play, suggest a marionette*" (Yeats 430).

Up to this point in the text, both the Musician's commentary and the author's textual notations (or "stage directions") seem consistent with a conventional (if abstract, or presentational) style of theatrical mimesis in which words stand in for objects and events not visible on stage. A somewhat different style of narrative accompaniment, however, begins to develop as the First Musician continues to observe and to comment on the Old Man. In the following speech the narrator's words begin to overlap the acts as they would be mimed by an actor; his commentary shifts from being mainly expository with respect to the events that are being enacted to apparently synchronous with them, so that the verbal account is thrown up against the actual sensory experience of the thing it is intended to represent:

> He has made a little heap of leaves;
> He lays the dry sticks on the leaves
> And, shivering with cold, he has taken up
> The fire-stick and socket from its hole.
> He whirls it round to get a flame;
> And now the dry sticks take the fire,
> And now the fire leaps up and shines
> Upon the hazels and the empty well. (Yeats 430)

This particular speech differs in important ways from the Fool's teichoscopy in *On Baile's Strand*. That speech was used principally so that the audience could picture an event otherwise difficult to stage—Cuchulain's attempt to slay the ocean; it works by retailing the action as if it were actually happening somewhere just out of sight. But here the musician's narrative is in some respects superfluous. The speaker describes in detail things the audience is seeing at the very same time they attend to his words; the aim in this case seems less to clarify the subject matter of the mimesis than to complicate or denature it. What happens as we hear it is surely a sense of awkwardness or impatience about listening to the narrative in the first place; in other words, the Musician's language "linearizes" an immediate sensory experience that is largely pictorial and simultaneous.

The estrangement that Yeats here achieves by juxtaposing scenic enactment with its verbal counterpart can be explained by reference to certain recent scientific research into the psychology of memory. Experiments have shown, for example, that verbal descriptions of something that has actually been seen—even a relatively uncomplicated thing such as a color swatch—demonstrably impair one's ability subsequently to recognize it (Gilbert 44–45). In an attempt to determine the relative accuracy of different ways of remembering sensory experience, the researchers J. W. Schooler and T. Y. Engstlere-Schooler asked subjects to recall both visual and verbal encounters with an object. The data were both disappointing (many of the subjects could not accurately identify a color swatch they had seen less than a minute beforehand) and also surprising: in apparent contradiction to the commonsense notion that seeing something provides for a fuller, more intense—and, therefore, more memorable—experience than hearing about it, researchers discovered that the verbal accounts of a thing almost always tended to overshadow actual memories of seeing it (Gilbert 271). Their conclusion was that the virtual "picture" created by the words was somehow interfering with the actual memory of seeing the object. In one study, for example, participants were given a color swatch and asked to study it with the purpose of remembering it. Half the group were then asked to describe the color, while half were not, and both groups were tested subsequently for their ability to pick out the original color from a lineup. Only one in three of the people who had had their visual experience confirmed in language proved able to pick the correct color, a task that was performed correctly by three in four of the "nondescribers." "Apparently," says Daniel Gilbert, "the describers' verbal descriptions of their experiences 'overwrote' their memories of the experiences themselves, and they ended up remembering not what they had experienced but what they had *said* about what they experienced" (Gilbert 44).

Experiments such as these have direct relevance to the affective experiences of the audience for Yeats' play. It is as if the Musician's words somehow cut in line, so to speak, tempting the brain into substituting an imagined picture for the real one. Each time the old man acts in such a way as to establish a coherent stage picture, his representation is displaced by the musician's speech, and this simultaneous imaginative framing of a theatrical act by its counterpart in narrative tends to cut against the grain of conventional mimesis. Mounding leaves, laying sticks, shivering, twirling a hardwood stick in a socket to generate heat, these are all actions readily imitable by any skilled performer, intelligible to audiences even without the use of props. But to have them narrated at the same time they are being acted gives the actor's mimesis an entirely different texture. The effect is like seeing labels—"book," "wall," "hat"—affixed prominently to the familiar objects one happens to be looking at. By having narrative accompany mimesis, Yeats writes into his play a kind of formal alienation effect: it is disconcerting for an audience to be instructed to imagine someone shivering with cold when they are fully aware that is, in fact, exactly what they are watching.

Yeats' narrator here does not substitute imagined activities for perceptual ones, as was the case with the classical messenger whose ecphrasis stands in for scenic enactment; instead, he lays one form of representation over the other. In Scarry's terms, a scene that is apprehended in an act of "immediate perception" is complicated by being made available simultaneously in words through an act of "mimetic perception." In effect, Yeats's dramaturgy requires an audience to imagine seeing things in the very act of seeing them. This sense of seeing something and being simultaneously asked to imagine seeing it is crucial to Yeats' refractive performance style. It is, literally, "a showing of showing," and it fosters the same kind of double consciousness that was an important component of Plato's advocacy of artistic transparency. The result is a characteristically modernist turn of theatre in the direction of mimesis mixed with diegesis. Like René Magritte's famous picture of a pipe that the artist cautioned was not a pipe, the musician's narrative qualifies the imitation of a thing with a reminder of its virtual status. His narrative deflects attention away from the thing imitated, toward the imitation itself and the signifying codes of its symbolic representation.

* * *

The hope that theatre might serve Apollo rather than Dionysos can be traced back to Plato and his call for a mixed mode of poetic performance. A performance entirely without mimesis would be ideal; but failing that, one ought at least to temper mimesis with diegesis. In the case of Yeats' theatre, what stands behind the tempering of mimesis is not mainly the older, moralizing strain of antitheatricalism; instead, Yeats' experiments with mediatized performance probably grow out of the broad abstractionist tradition in the arts originating at the end of the nineteenth century. In the visual arts, this rejection of the mimetic pictorial tradition most often took the form of an attempt to liberate the pictorial or sculptural object from any reference to a reality exterior to it. "It is well to remember that a picture," wrote the painter Maurice Denis in 1890, "before being a battle horse, a nude woman, or some anecdote—is essentially a plane surface covered with colors assembled in a certain order" (Chipp 94).

Yeats' motives in asking his audience to imagine the dramatic representation at the moment of its enactment are chiefly literary and aesthetic. *At the Hawk's Well* combines actors' mimetic performances, which require a degree of representation consistent with empathic involvement, with an abstract, objective aesthetics. But this diegetic method contains the seeds of a contradictory style of theatrical representation that can form the basis of an expressly political dramaturgy. Other modernist playwrights invent a mode of theatre in which the distancing effects of narrative are used to make dramatic performance into something quite different from the fully mimetic structure it normally is. This swerving of theatre away from mimesis can be documented early in several modernist writings on the subject. F. T. Marinetti, for example, called for separating theatre from "theatricality"(Puchner 7), and Brecht, in a similar revolutionary spirit, wanted to remake mimesis by incorporating elements that were foreign to it, in effect making theatre, whose essence is mimetic performance, into something it naturally wasn't. In a passage from Brecht's *Messingkauf Dialogues*, for example, a spokesman for conventionally mimetic performances ("the Dramaturg") cites Aristotle's *Poetics* in making his claim that a theatre without imitations "would no longer be theatre." (*The Messingkauf Dialogues* was left unfinished at the time of Brecht's death; cast in the form of a Platonic debate, its title, "The Brass-Buyer Dialogues," alludes to a person who is interested in theatre only for its base elements, as a buyer of scrap metal is interested only in the metal itself, not the particular ornamental or expressive forms into which it may have been shaped.) A theatre without mimesis might sound like an impoverishment, but for Brecht, as for Plato, it was precisely this kind of denatured or "untheatrical" theatre that was highly desirable. "Theatre" and "theatricality" are synonymous both with purely mimetic representations as well as with a kind of generalized perceptual numbness on the part of the spectators. Brecht's response to the Dramaturg, therefore, is spoken by a character who takes a more skeptical, intellectualized view of theatrical performance. A figure called "the Philosopher" admits that without mimesis there could be no theatre; but this state of affairs, for the Philosopher (and presumably for Brecht as well), seems not to pose a problem. The Philosopher's explanation of this apparent paradox is both serious and sarcastic:

> *The Dramaturg*: Imitations cut off from their purpose wouldn't be theatre, let me remind you.
> *The Philosopher*: That needn't particularly matter. We could call the result something different: "thaëter," for instance. (Brecht, *Messingkauf Dialogues* 16)

Like Yeats, Brecht made a bias against conventional mimesis into the cornerstone of his dramaturgy. But unlike Yeats, Brecht's prejudice against mimetic involvement is so extreme it is sometimes hard to distinguish it from older, cruder antitheatricalist rants. Brecht credited theatre absolutely with the power to bring about permanent behavioral or psychological change in its audience. "Let us go into one of these houses," he writes, describing a typical visit to the theatre, "and observe the effect which it has on the spectators". "Looking about us," Brecht says,

> we see somewhat motionless figures in a peculiar condition: they seem strenuously to be tensing all their muscles, except where these are flabby and exhausted. They scarcely communicate with each other; their relations are those of a lot of sleepers.... True, their eyes are open, but they stare rather than see, just as they listen rather than hear. They look at the stage as if in a trance: an expression which comes from the Middle Ages, the days of witches and priests. Seeing and hearing are activities, and can be pleasant ones, but these people seem relieved of activity and like men to whom something is being done. (Brecht, *Brecht on Theatre* 187)

The stage, Brecht insisted, had to be purged of everything "magical," by which he really did mean that an audience had to be prevented from a wholesale suspension of disbelief. The spectators Brecht envisioned were not to be coldly rational, however, in their response to mimetic performance, though his ideas are often so misconstrued. As Brecht outlines it, the ideal spectator's point of view seems less juridical than narratorial—quite similar, both in broad outline as well as in specific formal stipulations, to the artistic guidelines for mimesis recommended by Plato. "In short," Brecht writes, "the actor must remain a demonstrator; he must present the person demonstrated as a stranger, he must not suppress the '*he* did that, *he* said that' element in his performance" (Brecht, *Brecht on Theatre* 125).

Brecht's numerous writings on theatre constitute the most completely theorized historical application of diegesis as a necessary qualifier or antidote to mimesis. Yet Brecht's theatre is novel and at the same time wholly conventional, because nothing in his dramaturgy can be said to have been his own exclusive invention. All of the so-called alienation devices that Brecht touted as essential to the creation of a new, "epic" theatre have been occasional parts of dramatic performances ever since antiquity. These would include familiar diegetic elements such as choral singing, narrators' prologues and epilogues, as well as artificial performance elements usually identified with what is often called "presentational" theatre, such as self-conscious performance styles, montage (as in medieval "cycle dramas"), actors who speak out of character, and anti-illusionist scenography—in short and in sum, anything that Brecht saw in the traditions of theatrical performance as thwarting strictly mimetic representations. It is deeply ironic that Brecht brought into existence a distinctively "Brechtian" performance style—surely the most innovative and widely influential style of the entire twentieth century—simply by recycling the detritus of two millennia of histrionic performances: "If we understand the epic tendencies in drama to be those that encourage the development of a mediating communication system," writes Pfister, "then they were a recurrent feature of dramatic texts long before Brecht ever thought of them" (Pfister 71).

What *can* be attributed to Brecht, however, are two important and related ideas: first, the integrative theorizing of a kind of drama that was expressly and intrinsically "non-dramatic," or "non-Aristotelian," and second, the idea that drama (more specifically, dramatic performance) be viewed from a sociological rather than a strictly aesthetic point of view. This latter follows naturally from Brecht's interest in the psychological dynamics of mimetic performance. For Brecht, as for Plato, the "imitations" of an artist necessarily include a great many different kinds and degrees of repetitions and reproductions, certainly on the part of the poet or performer but also on the part of the viewing audience. These collective psychologies define a shifting range of mimetic possibilities for any play text. Dramatic performances in particular make use of familiar forms of conscious imitation in the theatre such as role-playing or scene-painting, but they also involve autonomic processes that bring together actors and their characters, or actors and their audiences, in unconscious identifications. Conventional dramatic performances make use of this multiplicity of imitations that are collected (to paraphrase Roland Barthes) not on the page nor even on the stage but in the spectator.

For much the same reasons as Plato preferred the narrator not to disguise the fact of his narrating, Brecht also favored "amateur" styles of performance that permitted spectators to see acting *as* acting. In his "Two Essays on Unprofessional Acting" (1940) as well as in the related piece written at about the same time, "Notes on the Folk Play," Brecht takes pains to dissociate his thinking from mere primitivism of the sort that had long been widely popular with modernist artists, especially among avant-garde painters and playwrights (Goldwater 15–43). "I do not by any means," Brecht wrote, "find simple acting *ipso facto* good or prefer it to anything less simple. I am not automatically moved by the enthusiasm of untrained or inadequately trained people who none the less feel passionately about art" (Brecht, *Brecht on Theatre* 148). And Brecht then goes on to describe his ideas for a theatre that was "naive but not primitive," one that, in other words, made a virtue of a shrewdly calculated amateurishness:

> And yet there is another kind of simplicity to be found in their [working-class] acting, a kind that does not result from a lack of origins but from a specific outlook and a specific concern. We speak of simplicity when complicated problems are so mastered as to make them easier to deal with and less difficult to grasp....This kind of simplicity does not involve poverty. Yet it is this that one finds in the playing of the best proletarian actors, whenever it is a question of portraying men's social life together. (Brecht, *Brecht on Theatre* 148)

Brecht's association with amateur actors seems to have been both fortuitous and fortunate. The *Lehrstücke* in particular seem to have been suited to non-professional performers: "his first acquaintance with 'proletarian actors' and singers," according to John Willett, "being evidently due to productions of *Die Massnahme* and *Die Mutter*" (Willett 152–53). Even the deprivations of the years he spent outside Germany and without a theatre may have contributed to the development of

Brecht's deceptively—but strategically—naive dramaturgy. Willett writes that

> [i]n exile he came into contract notably with the German semi-amateur groups in Paris who gave the premières of *Furcht und Elend des Dritten Reiches* and *Señora Carrar's Rifles* (1938 and 1937 respectively), with the New York Theater Union, Unity Theatre in London, the Copenhagen Revolutionary Theatre under Dagmar Andreassen and other Scandinavian amateur companies. (Willett 153)

It seems that for Brecht, an element of what comes across as a kind of professional naiveté (rather than ineptitude) on the part of the amateur actor is dramaturgically valuable , as was Socrates' disclaimer that he was not poetic; that is, in both cases a measure of apparent (whether calculated or accidental is immaterial) artistic disability is to be desired precisely because it throws the entire performance into critical relief. Conventional mimesis is rejected in favor of a productive, formally self-conscious, and partly diegetic style.

Much of Brecht's work first gained attention because of its contrast with then prevailing realistic styles of theatre; his polemics against "Aristotelian" drama are perhaps best understood as alternatives to the German theatrical tradition of the early decades of the twentieth century. Within that context, uncritical emotional involvement and an Aristotelian catharsis are among the targets Brecht frequently singles out for criticism, and his early, enthusiastic rejection of "empathic" experience in favor of a more contemplative outlook on the part of spectators has frequently been taken to represent the cornerstone of his theatre theory. But however undesirable wholesale emotional responses might have been to Brecht, equally objectionable was the realist mis-en-scène itself that Brecht and other modern playwrights inherited from the previous century. This stage tradition of "fourth wall" realism, with its naive and loving attention to the familiar details of everyday life, was in some respects as much a focus of Brecht's criticism as actors who vanished behind their roles.

Brecht's chief visual contribution to twentieth-century theatre in this respect was in his critical reformulation of the *Bühnenbild*, or "scenic picture," a trope that had governed European set design from roughly the middle of the eighteenth century when the painter Philip James de Loutherbourg collaborated with David Garrick to bring a new era of pictorial illusionism to their theatre at Drury Lane. For Brecht, however, the problem with conventional set design (or *Inszenierung*, "the scenicking," in the more precise Germanic term) was that it was understood to be merely supplemental to the action, or "story." Its chief function was to offer the audience a picturesque background or apparently natural environment against (or in) which events were seen to unfold.

In place of the *Bühnenbild*, Brecht and his long-time friend and scenographic collaborator, Caspar Neher, substituted the *Bühnenbau* ("stage-work" or "stage-construction"). Neher, who had known Brecht since the two were schoolmates before World War I, had been trained as an illustrator and painter; to Brecht's texts, beginning with early works such as *Baal* and *Drums in the Night* (1919–1922) and culminating with *The Threepenny Opera* (1928), their major collaborative effort, Neher contributed a series of drawings and sketches that were not so much designs for costumes or the set, or realizations of the text and, therefore, subordinated to the characters and action in an environmental sense, as they were (like the screen projections) a critical starting point, a kind of visualized rhetorical stance. Thus as Brecht described them in the case of the *Songspiel Mahagonny*, Neher's sketches were "quite as much an independent component of the opera as are [Kurt] Weill's music and the text" (Baugh 242). The set for a Brecht play, in other words, was intended to be conspicuously separable from the mimesis in much the same way as the actors were to be alienated or estranged from their roles by understanding them as it were from the perspective of one who tells their story.

As Brecht and Neher conceived of the *Bühnenbau*, the stage was not to be, as was the case with early modernist theatre artists such as Adolphe Appia, W. B. Yeats, or Gordon Craig, a work of art *en soi*, imitative of nothing other than itself. Rather Brecht was to perform with the stage picture the same radical surgery he had accomplished on the enactment of the text: it too was to be made more like narrative, mimesis transformed by rigorous application of diegesis. As early in his career as 1931, in commentary written subsequent to the performance of *The Threepenny Opera*, Brecht had called for the "literarization" of the theatre. At this point Brecht was still casting about for words to describe his own theatrical style, regularly using, in addition to the familiar term "epic" theatre, terms such as "dialectical" or "non-Aristotelian."

Like much of his criticism, Brecht's writings here seem to be a sequence of preliminary notes rather than a fully realized critical analysis. But central to his argument is the novel idea that by using specifically visual techniques borrowed from the then developing film industry, namely titles and screen projections, scenography itself might be made conspicuously "literary":

> The screens on which the titles of each scene are projected are a primitive attempt at literarizing the theatre. This literarization of the theatre needs to be developed to the utmost degree, as in general does the literarizing of all public occasions. Literarizing entails punctuating "representation" with "formulation." (Brecht, *Brecht on Theatre* 43)

Of course, as Baugh notes, the efficacy of Brecht's scenography depends to a great extent on its relative novelty, "its contrast with prevailing theater styles and audience expectations" (Baugh 250). Still the importance Brecht places on "punctuating" the various elements of theatrical performance, whether the actors' speaking of the text or the scenographic elements of the production, demonstrates the consistency of his general approach to theatre: mimesis was in every respect possible to be tempered with narrative forms. As for the effect of introducing materials extraneous to the mimesis—for example, giving individual scenes titles by using filmic projections or placards—Brecht concedes that "[t]he orthodox playwright's objection to the titles is that the dramatist ought to say everything that has to be said in the action, that the text must express everything within its own confines" (Brecht, *Brecht on Theatre* 44). Yet it was precisely this kind of diegetic overlay of mimetic performance that Brecht sought when he theorized about theatrical performance. His object was to turn the experience of mimesis partly into an experience of narrative, that is to say, into the mediated style of dramatic performance described by Plato. Play-watching was to be made a more "literary" activity, something resembling the act of play-reading. As Brecht puts it, "Footnotes, and the habit of turning back in order to check a point, need to be introduced into play-writing too" (Brecht, *Brecht on Theatre* 44).

For both Brecht and Neher, therefore, scenography was to be a conspicuous and important part of this newly "literarized" performance. In this endeavor, Brecht's *Inszenierung* became wholly consistent with other elements of his theatrical aesthetics. Just as Brecht intended the actors not to suppress the "he saids" or "she saids" of their performances, the set was itself also to contribute to the narrative style of the production. Hence Brecht's plays beginning at about this time could and did include, in addition to extrascenic objects that identified time and place, projections of newspaper cuttings, placards, music, songs, even, on occasion, direct exhortations to the audience to "follow the words in your programmes and sing along loudly!" (Baugh 240). This melange was to become the unique signature of "Brechtian" theatre, a visual style that as much as anything was aimed at destroying—or at least exposing—conventional pictorial concepts of the mimetically conceived "scene." As a result, the *Bühnenbau*—just like the dialogue or the actors' presentation of their characters—became part of the "fable," or diegesis, of the play.

If the look of the set was important for Brecht, the way the actors performed their roles was even more so. Brecht admired oriental acting styles (or what he perceived to be oriental acting styles); his description of the "splendid remoteness" of Chinese acting in its broad outlines resembles the strangeness that Yeats valued in his own productions. It is likely that the "strangeness" both Brecht and Yeats saw in Asian performances was due mainly to their relative ignorance of those theatres. Yet rightly or wrongly, Brecht, like Yeats, found in oriental acting a means partly to disengage the actor from the role and the audience, in turn, from fully empathic involvement with the represented character. His immediate aim was not so much to banish mimesis—without imitation, as Brecht certainly knew, there is no theatre—as to ensure that actors' imitations were never unqualified by cognitive awareness *of* mimesis. Rather, Brecht sought to include in the theatrical performance, as Herbert Blau puts it in contemporary theoretical terms, "a *productive emptiness* which, as it throws the elements of presence into relief, the lineaments of the representational structure, puts into question the (falsifying?) distance between representation and nonrepresentation. Not a void, then, but a *disequilibrium*—which is the space of contradictions" (Blau 192).

It was the apparent property of self-awareness that makes the Chinese actor so attractive to Brecht, just as it was the specific rhetorical consciousness of the narrator's stance with respect to his material that so intrigued Plato. This is one of the main reasons that in some of his early works Brecht experimented with narrators or narrator-figures whose purpose was to frame the mimesis conspicuously as an enactment. The *Lehrstücke*, for example, include narrative choruses who summarize events and introduce characters: the "Control Chorus" in *The Measures Taken*, or the Great Choruses in *He Who Says Yes* and *He Who Says No*. Central to the development of Brecht's thinking on theatre as well as to his overall dramatic aesthetic, the *Lehrstücke*, a group of nine plays written between 1926 and 1933 (and written at least partly in collaboration with Elisabeth Hauptmann), were so-named to distinguish their dramaturgy from that of Brecht's other, more commercially successful, plays as well as from conventional theatre in general. In writing about them, Brecht used the term "major pedagogy" (as distinguished from the "minor" pedagogy of the "epic theatre") to identify a wholesale reformation of the relationship (or "traffic") between actors and audience. According to the major pedagogy of theatre, spectators were no longer to be passive consumers of mimesis, even if enlightened or "alienated" with regard to the actions performed on stage. Instead, all distinctions between the stage and the audience were to be abolished. The terms "actors" and "audience" would be rendered technically obsolete, and both groups of people would instead collaborate in the mounting of a performance of the text. Thus, for example, Brecht envisioned his play about Lindbergh's flight as suitable for enactment as a schoolroom project, and a performance of *The Measures Taken* in Berlin in 1930 "cast" several thousand local workers in the role of the Central Chorus.

Because of their heavy investments in bringing about actual political change, the *Lehrstücke* have in the past been taken to represent Brecht's "vulgar Marxist" phase. But since Rainer Steinweg's important study of 1972, these plays have been seen increasingly as central to Brecht's theory and practice of theatre (Wright 14). As a group, the *Lehrstücke* include, in addition to narrators, many instances in which the characters break frame to describe to the audience the actions that they are about to represent. Here too the point is to balance mimetic enactment with diegetic materials or frameworks. In one of the best known of the *Lehrstücke*, *The Measures Taken*, most scenes begin with an expository narrative. At the beginning of the first scene, for example, four communist activists, performing for the most part like a chorus, speak what seems to be a kind of prologue to the central mimesis. The group identify themselves, their situation, and the action to come:

> *The Four Agitators*: We came as agitators from Moscow. We were to proceed to the city of Mukden for propaganda purposes and to aid the Chinese Party. We were to report to the Party House, the last before the border, and ask for a leader. We were met there by a young comrade and spoke of our mission. We repeat the discussion. (Brecht, *The Measures Taken* 9)

There is one important difference, however, between Brecht's exposition and more conventional uses of narrative to provide the background for dramatic action. In addition to providing the audience with information about the characters, setting, and plot, the prologue concludes with the speaker asserting that the enactment that follows on stage will "repeat" certain past events. His insistence on a dramaturgy of repetition is crucial to the design of *The Measures Taken*. What Brecht is striving to achieve here is to replace conventional theatrical mimesis of events as if they were occurring for the first time with a conspicuously post-hoc enactment that is being reconstructed from memory of events that are understood to have already occurred. The purpose is to orient spectators to engage with the represented events as a kind of history, which is to say that Brecht wants spectators not to blur the distinction between a staged enactment and simultaneously occurring action. In this project, narrative plays a key role, not so much to make absent events "present" as to represent them in a "literary" way, to present them as occurring in the mode of "storytelling." On this stage, mimesis is strategically separated off from the "reality illusion" that normally characterizes theatrical performance and located instead within a larger diegetic structure.

The Measures Taken dramatizes events reenacted for the express purpose of demonstrating history. The play tells the story of a group of five communist agitators, four men and one woman, who travel from Moscow to a Chinese city to spread propaganda surreptitiously among the local workers. It is essential to their success that they disguise themselves and follow strictly the instructions they have been given by party superiors. But one of the agitators is young and impatient with party discipline and its methods that he cannot understand. Because of his inexperience, and particularly because he cannot suppress his identity and his anger when he encounters people who are suffering under despotic rule, he inadvertently reveals who he is and puts everyone else and their entire mission at risk. The agitators call a hasty meeting to discuss their options, and all agree that they cannot hope to escape over the border so long as their young comrade is with them. But neither can they simply leave him behind: he would surely be caught and executed, and since he would be recognized the work of the group would be betrayed. After considerable debate, the four agitators decide to shoot their comrade themselves and cast his body into a lime pit where the caustic substance would burn away any traces of his existence. This they do; then they return to Moscow where they explain to Party superiors what has happened. Their tale of events, part enactment and part narration, make up the dramatic structure of *The Measures Taken*.

Brecht's plot is grim but by no means unrealistic, especially given its historical context within Europe in the early 1930s. But the most shocking aspect of Brecht's play is that the young comrade himself takes part in the discussion in which his fate is decided, and, in what is probably the most politically controversial aspect of the drama, he acknowledges the correctness of the other activists' decision. Even though the young comrade's sacrifice for the greater good of his friends and their mission differs in no fundamental way from the noble, selfless act of a combat soldier who saves the rest of his squad by giving up his own life, the scene nevertheless proves distressing to most audiences. The activists' decision to shoot their comrade seems ruthless, all the more so because of the surprising willingness of the victim to acknowledge the rightness of his comrades' judgment. As Brecht's actors recapitulate the scene, the decision to murder the young comrade is made to seem the product of a completely dispassionate rationality:

> *The First Agitator* to the young comrade: If you are caught you will be shot; and since you will be recognized, our work will have been betrayed. Therefore we must be the ones to shoot you and cast you into the lime-pit, so that the lime will burn away all traces of you. And yet we ask you: Do you know any way out?
> *The Young Comrade*: No.
> *The Three Agitators*: And we ask you: Do you agree with us?
> *Pause.*
> *The Young Comrade*: Yes.
> *The Three Agitators*: We also ask you: What shall we do with your body?
> *The Young Comrade*: You must cast me into the lime-pit, he said.
> (Brecht, *The Measures Taken* 33)

The assassination itself is carried out with compassion, a compassion that seems all the more shocking because it seems so incompatible with murder:

> *The Young Comrade*: Help me.
> *The Three Agitators*: Rest your head on our arm. Close your eyes.
> (Brecht, *The Measures Taken* 34)

Following the dramatization, the members of the tribunal that has been convened to investigate the assassination conclude that the agitators acted properly in carrying out "the teachings of the Classics, the ABC of Communism."

Two decades after the wholesale collapse of communism in Europe, the Marxist dogma expressed by Brecht's characters sounds quaint, if not downright ludicrous; it's hard to believe somebody as politically savvy as Brecht could have been so naive. More troubling to many commentators is that Brecht seems to have been willing to excuse even Stalin and his mass persecutions so long as they advance the Communist cause; read as an apologia for Stalin's brutal purges, *The Measures Taken* becomes frankly appalling:

> Sink in filth
> Embrace the butcher, but
> Change the world: It needs it! (Brecht, *The Measures Taken* 25)

The Measures Taken, like most (if not all) of the *Lehrstücke*, is sometimes classified as a "thesis play," which it indeed resembles, though to be sure the particular "thesis" that Brecht sought to advance seems to vary widely according to the politics of the commentator. "[T]he humanists," writes Elizabeth Wright, "thought that Brecht was being Leninist and the Marxists feared that the play would be seen as Leninist by the humanists" (Wright 16). More recently this and other *Lehrstücke* have been associated not so much with communist dogma as with Brecht's experiments with the relatively spare aesthetics of Asian theatre (notably the Japanese Noh drama *Taniko*, which he had read in Elisabeth Hauptmann's translation from Arthur Waley's *The No-Plays of Japan*). The *Lehrstücke* have been linked as well to Brecht's interest in broader cultural problems concerning the "post-humanist or post-individualist subject" (Jameson 62). I would add only that from a purely technical point of view, the most interesting feature of these plays is their insistent "undramatic" quality. In the *Lehrstücke* Brecht seems to have turned to a primitivist dramaturgy in an attempt to see the extent to which theatre could be "literarized" or even turned into something it was naturally not.

Viewed in this way as an experiment in "thaeter," *The Measures Taken* is not narrowly propagandistic (nor even didactic, at least as that term is commonly understood); rather its dramaturgy can be characterized as broadly narratorial, less mimesis than *histoire*. "The whole point of history," says Arthur Danto, "is *not* to know about actions as witnesses might, but as historians do, in connection with later events and as part of temporal wholes. History, in other words, only emerges once the game is over" (Danto 183). Like the other plays of its kind, *The Measures Taken* remains, in more ways than one, as Roswitha Mueller calls it, "a genuinely utopian project" (Mueller 90). The purpose of the narrative displacement is to render the action in a relational rather than substantive way, to highlight the fact of the actors' separation from actual events. Those events are not seen by spectators as if they were witnesses. They are instead re-produced. Like a conventional literary narrative, Brecht's play is not so much an imitation as a presentation. All the chief episodes and incidents are understood to be secondary to the recounting of them before witnesses, as if the act of telling were important in its own right. Brecht's stage does not attempt to bring unmediated experience before the eye, rather to remember it.

The *Lehrstücke* represent Brecht's most emphatic break with conventional theatre, but even in his less experimental works he substitutes conventions of storytelling for more direct styles of representation. Take the scene headings that are projected onto a screen during a production of *Mother Courage*: for example, "Mother Courage sings the Song of the Grand Capitulation" (Scene 4) or "Mother Courage at the peak of her business career" (Scene 7). These textual summaries "literarize" the action that is being played out before spectators; they are a way of holding the performance constantly at some distance or at least undercutting it, putting it, as it were, in quotes. Because these narrative summaries are in view throughout any given scene, the enactment remains conspicuously partial. To have in full view a text such as "Mother Courage Sings the Song of the Great Capitulation," at the same time an actor sings a song by that name, "estranges" the performance in much the same way that the running subtitles in a foreign film compel one to consider simultaneously both the action and its symbolic representation. Text and enactment together set up a kind of visual equipoise, so that the performance makes constant reference to something that is not directly accessible.

In this sense Brechtian performance is almost always a dramaturgy that is calculated to demystify an audience. This is true even of those moments when the action is most fraught with emotion. The scene in which Kattrin is executed while drumming to warn the townspeople of Halle of an imminent attack, and Helene Weigel's agonizing "silent scream" over the death of her son Swiss Cheese are surely two examples of the most moving episodes in all of twentieth-century theatre. Both represent moments in which the staged events are most likely to seem the equivalent of raw, sensual experience. "In view of this," Brecht wrote in his notes to the scene with Kattrin on the roof, "the actors in rehearsal were made to add 'said the man' or 'said the woman' after each speech" (Brecht, *Mother Courage* 138). Although such rehearsal techniques have not historically kept audiences from being swept away by Kattrin's heroism, or by Anna Feierling's losses, they at least testify to the overall consistency of Brecht's method. Not that actions on Brecht's stage are not rendered faithfully as if from life—Brecht once insisted during rehearsals for a production of *Edward II* that the actors who performed a hanging were to become so thoroughly knowledgeable about that occupation as to become "virtuosos of the gallows." But the various conventions that have come to be associated with epic theatre—songs, placards, filmic projections, the A-effect, direct address, *gestus*—these all contribute to a Platonic ideal for mimesis, to a theatre that is openly narrative and "literary" at the same time it is "dramatic."

* * *

Brecht leads the way, both in his theory as well as in his practice, for a new kind of theatre that was to be formally estranged from straightforward enactment. Numerous other dramatists from the last half of the twentieth century have similar ambitions, with respect to both theatre and politics, and a good deal of the drama of that period (as well as some of its cinema) owes heavy debts to Brecht's work. Postwar British dramatists especially are fond of adding an "epic" signature to their plays, typically—following Brecht's lead—for political rather than aesthetic purposes. Like Caryl Churchill, Pam Gems structures her plays to include openly diegetic markers. In *Piaf* (1978), for example, Gems interlaces her biography of the famous French singer with music from a number of Piaf's recordings; often these are not part of the enactment but come from the offstage, as when the music "Un sale petit brouillard" is heard while Piaf is seen on stage "getting it from a Legionnaire" (Gems 16).

Diegetic interventions in Gems' play like this one supplement those moments when the character Piaf (played in the original production by Jane Lapotaire) herself is seen singing. Both kinds of songs frequently impede the progress of events: either they interrupt the sequence of enacted vignettes, framing them as gestural moments in the manner of Brecht, or they set up an ironic tension with the representation whenever "real" music is heard outside the dramatized world of the play.

Some early reviewers of Gems' play were generally critical of Gems' dramaturgy—Frank Rich of the *New York Times* called the play "slight" and complained that it was full of "sketchy" people—but from a distance of thirty years her artistry seems more calculated and successful. In documenting the seamy side of the life of France's most famous singer, Gems attempts more than to rescue the "real" Piaf from the legends that have become attached to her. For one thing, Gems shows the extent to which the identity of this woman is not her own but something constructed for her. For example, early in the play, Louis Leplée sees the talent in a young woman named Edith Gassion, who sings, in the club he owns, as an opportunity to make money. Over her objections, he bestows on her a stage name: Edith Piaf. And Gems' drama also frames important questions about the nature of the relationship between art and the artist. Contrary to romanticized or heroic notions of artistic genius, Gems's Piaf seems not to be motivated by anything having to do with her individual psychology or personal biography; she does not sing out of sorrow, or need, not even for the love of singing itself. Rather, in dramatizing the biography of Edith Piaf, Gems depicts only an incidental or accidental connection between the person and the art. Her viewpoint is both at once postmodern and ancient, a portrait of the artist according to Barthes as well as Plato, both of whom believe the author to be merely the purveyor of a "text" over which he or she has little actual control.

The songs and the episodic plot in *Piaf* are emphatically Brechtian, as are some of the cartoon-like characters: these include a pair of drunken American sailors, for example, who proposition Piaf—successfully—in a bar and then are never seen or heard again, or two sterotyped Germans who click their heels and speak their clichés in thick, Teutonic accents. The oddly incomplete realizations of these figures recall the "sketchiness" typical of some of the earlier Brecht-Neher collaborations. In working out the look of a Brechtian performance, for example, Neher, according to Baugh, seemed to be "trying to find a theatre equivalent of the sketch: a way of bestowing wood, canvas and stage paint with a softness of definition similar to the undogmatic, thought-provoking effects achieved by drawing with ink upon damp watercolour washes, a favoured medium at this time" (Baugh 236). Brecht sought in his productions to answer naturalism with a conspicuously stylized mimesis, and Gems' approach to these minor characters seems calculated to have a similar antiphonal effect. Their radical simplification seems to be an artistic element in its own right; if the multitude of "sketchy" characters do not seem fully to belong to the dramatized world, it is because they are intended partly to place the mimesis in quotes.

A Platonic skepticism with regard to theatrical representation can be found also in the work of the British "fringe" playwright Howard Brenton. Writing drama, as he put it in an interview, "unreservedly in the cause of socialism" (Carlson 478), Brenton takes from Jerzy Grotowski the notion of a "poor" theatre and gives it an expressly political twist. In the introduction to his collection *Plays for the Poor Theatre* (1980), Brenton says that "[t]hese five plays in varied ways try to turn 'bad theatrical conditions' to advantage. They are not easy to do or constricted in what they say—their 'poverty' is that of theatre companies with no money, amateur acting, touring conditions that can vary from a studio theatre to a school gymnasium, to a room with a bare floor and no electric plug."

One could think of this fondness for inexpensive productions as squarely in the tradition of Brecht's essays on "unprofessional" acting or the shrewd disavowal of mimesis expressed by Socrates in *The Republic*. That is to say, Brenton's poor theatre is not really "theatre" in the way that Plato's unmetrical recounting of events is not "poetic." As was the case with Socrates' presentation of an episode from the Trojan War "without imitation," Brenton's works for the stage make deliberate use of non-mimetic elements. The short play *The Saliva Milkshake* (1975), for example, is constructed according to a classical economy of representation with respect to the placement of violence off stage. The play dramatizes the situation of a man caught between personal and public loyalties. Martin and Joan (Brenton gives us only their first names) were close friends when they were graduate students in Britain in the 1960s; since then they have had no contact. Martin is now a research scientist working for a British university. Because of the high quality of his research, he is a candidate for an important academic chair in biological science. Joan, on the other hand, belongs to a group of left-wing terrorists. She has just shot and killed the home secretary, and she comes to Martin for temporary asylum and to ask him to fetch her a new passport. Without warning, Martin is thrust into a situation in which he must choose to betray either his country or his old friend. The play includes a great deal of violence: the assassination of the home secretary is described by Joan in some detail; Joan, fearing capture, eventually turns her gun on herself; and Martin has his eardrums punctured by the terrorists in retaliation for betraying their plot to the government's Special Branch.

But none of these acts takes place on stage. The play features a kind of cinematic voice-over during most of its scenes. Brenton's drama blends narrative, exposition, and simplistic enactment in a way that recalls the style of some of Brecht's *Lehrstücke*; the actor playing Martin especially steps in and out of character and in and out of scenes over the course of the play, explaining events, setting the action in context, and commenting on his own involvement. At one point during their first conversation Joan silently mouths a revolutionary slogan while Martin speaks a narrative aside; at such moments of overlap, the play incorporates both mimesis and diegesis:

> She spat a little. A little saliva. I realised how wet the inside of her lip was. How close her teeth were and impervious to the juices of her mouth...That in her chin there must be a runnel...Below her lower lip...Gutter, lake, swamp, with the juices...Swishing...I became obsessed with this trivial, physical particularity...Her wet mouth...I couldn't follow what she was saying. I wanted to take a straw, milk straw, and suck out the wet. I fantasised on this. She was speaking of England and Ireland...Terror, abuse of working people...The long warp of religion and class warfare and...I could only think of her mouth. Little pumps, irrigation systems. A saliva milkshake. I'm not a political person. (Brenton 13)

In Brenton's theatre, the limited finances of a fringe acting company become the basis of a virtuous aesthetic. A company with no money to build a set representing "Euston Station," for example, must "stage" that environment with narrative alone. But that narrative could also be made to function as part of a coherent and constructively artificial theatrical style. The way to deal with a lack of money for sets, in other words, is to invent a dramaturgy independent of such sets in the first place. Bare stage and narrative then become part of an obviously diegetic style of presentation, one that depends as much on the imaginations of spectators as on conventionally mimetic (and more expensive) enactments.

Speeches such as the foregoing in *The Saliva Milkshake* in which a character pauses during the action to speak his or her private feeling are fairly common among plays of the twentieth century. The most extreme example among modern plays would doubtless be Eugene O'Neill's *Strange Interlude*, where the majority of speeches over the course of an entire trilogy are understood by convention to represent the different characters' unspoken thoughts. But the technique is much older, and to see the relative novelty of Brenton's dramaturgy it will be useful to distinguish Martin's discourse from monologues used by earlier playwrights in their attempts to dramatize a character's innermost thoughts. One of the earliest instances of this kind of narrative aside occurs in George Lillo's sentimental tragedy of the middle class, *The London Merchant* (1728). Lillo's play shows how George Barnwell, a good-hearted but naive youth, is corrupted by the schemes of a wicked woman named Millwood. Several times during the course of the play, the action seems to develop on two levels simultaneously as characters speak first to one another and then to themselves. Soon after Barnwell meets Millwood (who is described by Lillo as "a lady of pleasure"), she lays her hand on his "as by accident." Young Barnwell is too flustered and inexperienced to frame a public response, but inside him an emotional storm rages. The text reveals this inner reality as a narrative aside: "Her disorder is so great she don't perceive she has laid her hand on mine. Heaven! How she trembles! What can this mean?" (Lillo 603).

The function of this kind of narrative is to make apparent to an audience aspects of character or mentality that are otherwise difficult if not impossible to stage. Not really an aside (it seems not to be spoken in collusion with the audience), nor a soliloquy in the conventional sense of the term, Barnwell's speech more closely resembles an interior monologue. Lillo interrupts the action at precisely those moments when he wants to define for spectators a crucial difference between the public, and superficial, aspects of young Barnwell's character and those thoughts or emotions that are inner and normally hidden, therefore, presumably more genuine.

As Brenton uses the device in *The Saliva Milkshake*, however, it is not only, and perhaps not even mainly, intended to publicize his character's private thoughts. Martin's commentary differs from young Barnwell's in that it frames the represented moment as if it were already history. Brenton shifts emphasis from a series of events unfolding in the present to something resembling a completed ideational sequence. In effect, Martin's aside makes drama into a dramatization. Without breaking frame by seeming to acknowledge the presence of an audience, as do some of the characters in Brecht's *Lehrstücke*, Martin's performance is nevertheless invested with what Brecht had called "a definite gest of showing" (Brecht, *Brecht on Theatre* 136). By superimposing narrative onto the represented events, Martin creates a discordant mix of past and present in which the speaker is seen to reflect on his encounter with Joan at precisely the same time he is seen to be experiencing it. To hear the history of their encounter told even as it is enacted before the audience creates a temporal ambiguity. It is as though the woman, while occupying the stage in what we would conventionally take to be real time and space, assumes simultaneously the status of a distant imagining because the speaker's narrative intrusion flatly contradicts the dramatic representation. Suzanne Langer, in her memorable definition of the essence of the different literary genres, called drama the coherent illusion of a "virtual future"; "it is only a present filled with its own future," she says, "which is really dramatic" (Langer 307). But Martin, stepping outside the plot and compromising the dramatic illusion with narrative, instead shows us that illusion's provisional substance.

* * *

The theories of Brecht are as much postmodern as they are modernist, as Elizabeth Wright (and others) have recently shown. "[T]he famous estrangement effect (*Verfremdungseffekt*)," she says, "the gestic style, the appeal to the spectator, may be seen as symbolic devices designed to disrupt the imaginary unity between producer and text, actor and role, and spectator and stage, an enterprise which is similar in spirit to Barthes's project in *S/Z* (1975) and A *Lover's Discourse* (1978)" (Wright 2). Wright's analysis suggests that, at least as far as drama is concerned, to classify Brecht, whose literary career bridged modernism and postmodernism, as belonging more to one era or the other is more illusory than real. And Wright's argument can be extended to include other theatre artists, calling into question what has sometimes been characterized as an aesthetic (or even ideological) "struggle" between modernist and postmodernist playwrights. A kaleidoscope might be a more appropriate metaphor to describe the history of drama in the twentieth century, as the relationship between writers who are chronologically either "modernist" and "postmodernist" often displays more of an essential kinship than a narrative of struggle and revolt against a precursor. Gordon Craig in the early years of the twentieth century was advocating much the same kinds of depersonalized stage figurations that Beckett was creating well into the 1980s in the short, haunting works termed "dramaticules," just as Brecht's calling in 1931 for a "literary" or "non-Aristotelian" mimesis—that is to say, a theatre deformed or partly dismantled by diegesis—has had wide application to much drama of the latter part of the twentieth century. In both cases the core aesthetic values, values that may be described collectively in terms of a broadly Platonic ideal for theatrical performance, point to a common historical style that links playwrights as different as Yeats and Brecht or Gertrude Stein and Samuel Beckett. In the dramas of all these writers, many of the familiar distinctions between "modernist" and "postmodern" styles are often hard to see or simply vanish. "A cursory look at the dramatic literature of the last decades," writes Puchner, "reveals the continuing persistence of many features that characterize the modernist diegetic theater" (Puchner 173). This suggests in turn that "Platonic" models for theatre may well represent something like a period style for the twentieth century. Two final variants of this partly diegetic theatre can be found in the work of Gertrude Stein and Ntozake Shange.

Stein's work for the theatre has always been problematic; she wrote for the stage more or less continuously for more than thirty years, and her status as one of the great literary modernists is secure. ("[I]t is indisputable," writes Bonnie Marranca, "that Gertrude Stein is the great American modernist mind" [Marranca *xxi*]). Yet in comparison with the other important figures of twentieth-century theatre, there is little commentary—and even less production history—to put her work for the stage in context. Her plays are never found in drama anthologies, and even theatre historians regularly pass her over. In writing about European drama between the wars, for example, Oscar Brockett in his influential *History of the Theatre* describes all-but-forgotten plays by Luigi Chiarelli, Tristan Tzara, and Fritz von Unruh. But Stein doesn't even get a mention in his index.

Part of the reason for the relative obscurity of Stein's plays is their notorious difficulty; as Edmund Wilson long ago observed, speaking with sarcasm but not without truthfulness or some measure of regret, "most of what Miss Stein publishes nowadays must apparently remain absolutely unintelligible even to a sympathetic reader" (Wilson 243). Indeed the texts that Stein calls "plays" seem to bear little, if any, resemblance to any other members of the genre. In speaking of her plays, Stein never used conventional descriptors such as plot, character, or action; her plays did not tell stories, neither were they imitations of experience. She commonly referred to her works for the stage as "landscapes," and those works could, accordingly, include dramatizations of countries ("Mexico") or states of mind ("Pink Melon Joy") as well as conversations or events from the lives of people. In comparison with Stein's far-ranging experiments with the genre, even the strange, late works of Beckett seem tamely neoclassical. One of the hallmarks of Beckett's theatre is his deployment of what has been called "non-scenic space," but the majority of Stein's plays seem incompatible with any actual depiction of exterior or interior "space" whatsoever. When her plays are staged, as the author of one recent survey concludes, "even [the] most imaginative and convention-defying of theatre artists had to fill some of her scenic absences in order to bring the texts to life on the stage" (Durham 4).

Stein's unusual theatre pieces are perhaps best understood as belonging to the theatrical avant garde, which throughout the twentieth century continued to display a particularly hostile attitude toward conventional drama. For Stein, as for Beckett, this resistance to theatre took the form of an almost complete break with any theatrical tradition or convention in order to dig down through the genre and its history to reach bedrock. It was only after mid-century (more than a decade after her death) that Stein's plays began to accumulate anything like a history of performance, as if (as sometimes happens in theatre history) they had to wait until such time as there developed audiences to appreciate them and styles in which they could be successfully performed.

Often those styles blend genres or modes: a piece such as *Photograph* (1920), for example, seems to inhabit the margins between prose and poetry or between narrative and drama. The play is subtitled "A Play in Five Acts," but only Act I has anything like the substantial story line or dialogic interactions between people that are characteristic of the formal structure of Western drama. Acts II and III of Stein's play, for example, are random snippets of commentary and/or conversation. Stein gives no indication as to who is speaking these words, or whether there is more than one speaker, or, for that matter, whether there are, in fact, any people at all present on stage:

Act Second
Two authors. Rabbits are eaten,
                    Dogs eat rabbits.
Snails eat leaves.
Expression falters.
Wild flowers drink.
The Star Spangled banner.
Read the notices.
Act III
A photograph. A photograph of a number of people if each one of them is reproduces if two have a baby if both the babies are boys what is the name of the street.
Madame. (Stein, *Last Operas and Plays* 154)

Act IV is similar in composition and length to Act II, and Act V verges on parodic nonsense. It consists of three cryptic sentences spoken by an unknown person about someone who has never been mentioned before: "I make a sentence in Vincennes. It is this. I will never reason away George" (Stein 154).

In *Photograph*, Stein specifies neither time, place, nor dramatis personae. The piece seems to record a chorus of different and unidentified voices speaking in different and unidentified circumstances; even the relatively lengthy first "act" of the work seems to lack a center, both thematically as well as dramaturgically:

> For a photograph we need a wall.
> Star gazing.
> Photographs are small. They reproduce well.
> I enlarge better.
> Don't say that practically.
> And so we resist.
> We miss stones.
>   Now we sing.
> St. Cloud and you.
> Saint Cloud and loud.
> I sing you sing, birthday songs tulip belongs to red cream
> and green and crimson so that the house chosen has a soft wall.
> Oh come and believe me oh come and believe me to-day oh
> come and believe me oh come just for one minute
> Age makes no difference. (Stein 152)

And so on for another page and a half of text. It is not at all clear to a would-be actor how a script such as *Photograph* is to be performed. For one thing, in the absence of explicit authorial instructions as to production (such as those Beckett almost invariably provides, for example, in the copious stage notes that accompany his late short works for the stage), it is impossible with Stein's text to answer some of the most basic questions of theatrical performance. There is no way to determine who is saying what, what is a stage direction, or what is a subjective response from a character, or even who or what is to be considered a "character" in the first place. Her play (to paraphrase Mark Twain's assessment of James Fenimore Cooper's "literary offenses") has no plot, no story, no order, no system or sequence, no seeming imitation of reality, and its language (as Stein's critics never seem to tire of complaining) treats words as if they had neither history nor content. As for other kinds of evidence of "authorial intent," the overt structure of the piece seems to be ironic or subversive, or, possibly, self-consciously "playful." (In subtitling her work "a play in five acts," Stein is surely acknowledging Horace's well-known dictum that a play should have "five acts, no more or less.") Indeed Stein's "play" seems intended to carry out a kind of autopsy on the genre, for it thoroughly dismantles all of the conventional markers of dramatic form. "Stein never stopped asking questions of dramatic form," writes Bonnie Marranca; "she wrote plays that consist of lists, objects, letters, sentences, and aphorisms. Their characters include cities, circles, religions, mountains" (Marranca *xii*). In fact, much of Stein's work for the theatre is so ideationally extravagant that little of it is possible to stage mimetically. Performance, therefore, requires the viewer to half-create what the minimal enactment does not supply. It is a drama of parts, of fleeting images and snatches of conversation; it is drama made of fragments not shaped into a unified whole with (as Aristotle prescribed) "a beginning, a middle, and an end."

Yet despite its wholesale destruction of mimesis, Stein's piece, like some of Beckett's ephemeral non-plays (the so-called dramaticules) or his mysterious narrative fragments and "fizzles," seems inherently "theatrical." If given a live performance, *Photograph* acquires the immediacy of the spoken voice that makes it accessible emotionally (if not immediately intellectually) to an audience. When they are embodied by actors, Stein's words, despite their dehumanized appearance, seem to acquire a direct and curiously dramatic quality. For example, in his 1977 production of *Photograph*, James Lapine's Performance Group (Open Space Theater, New York) assigned the text to eight different actors—five women, two men, and one child. Some of the lines were narrated, some spoken by the performers as if in character (though the text itself gives no indications whatsoever as regards time, place, or even the number or sex of the speakers). Lacking all traces of "plot," "character," and "action," the play is expressly "non-Aristotelian." It would hardly be possible for actors to identify with the roles they had been assigned or to psychologize their "characters" to any significant extent. In Lapine's performance, individual subjectivity is blanked out in service of the delivery of the text; the actors speak more in the manner of ventriloquists than characters in whom the words originate, and the overall aesthetic effect is closer to a stage reading. "Stein's plays," says Jane Bowers, "oppose the physicality of performance. Stein's is a theater of language: her plays are adamantly and self-consciously 'literary'" (Bowers 2). That is to say, if performances of Stein's plays have often proved to be theatrically effective (as repeatedly they have), their success on stage is due less to their inherent theatricality (as that term is understood in the conventional sense) rather than to the general willingness of present-day audiences to consider any kind of public recitation itself an intrinsically "dramatic" act. One could say that for contemporary audiences who have had widespread exposure to Brechtian mediated (or "literary") performance styles, the staged reading has itself become a legitimate mode of theatrical performance. In such cases the visible interplay between actor and his or her role/text becomes every bit as entertaining and intellectually substantive as more conventional mimetic enactments.

A Play Called Not and Now (1936), one of Stein's later works for the theatre, illustrates her continued interest in creating a mediatized stage, one in which a diegetic framework is made a conspicuous part of the theatrical performance. Stein wrote this much longer piece (also composed, like *Photograph*, in "acts") shortly after attending a celebrity event in Beverly Hills in 1936; the piece was first performed more than fifty years later, when Hanne Tierney gave it a one-week production run at the Choice Theater in New York (March 10–17, 1989). Stein's work makes use of some of the traditional mimetic structure of the genre, depicting recognizable persons—Hollywood celebrities of the day, for the most part—in common social situations. But these mimetic elements are made subject throughout the work to a generalized narrative consciousness. For example, in a gesture that recalls Brecht's definition of the actor as "demonstrator," Stein's text begins by identifying her characters as fictions that are produced by actors:

Characters
A man who looks like Dashiell Hammett
A man who looks like Picasso
A man who looks like Charlie Chaplin
A man who looks like Lord Berners
and a man who looks like David Green.
Women
A woman who looks like Anita Loos
A woman who looks like Gertrude Atherton
A woman who looks like Lady Diana Gray
A woman who looks like Katherine Cornell
A woman who looks like Daisy Fellowes
A woman who looks like Mrs. Andrew Greene.
(Stein, *Last Operas and Plays* 422)

Normally audiences get to know the characters in a play serially, as each in turn comes on stage; preliminary lists of dramatis personae are included in published versions of plays mainly as a convenience to readers. But Stein's play seems to be organized so as to make this textual element part of the performance, in effect "literarizing" it in something like the manner prescribed for performance by Brecht. The play indeed seems to require a narrative voice to present this list of characters and then to set the drama in motion; after hearing the above list of dramatis personae, for example, the audience is told that

> These are the characters and this is what they do.
> A man who looks like Doctor Gidon and some one who looks like each one of the other characters.
> The play will now begin. (Stein 422)

Stein's text maintains this narrative voice more or less constantly throughout the play. Each act begins with a narrative overview, as, for example, Acts I and II begin, respectively, with the following diegetic commentary:

> Act I
> The ones who look like Dashiell Hammett Picasso Charlie Chaplin and Lord Berners stand around....
> Act II
> Now comes the time when they come in one by one, they are not alone as they come in one by one because there is never any other one. (Stein 423–28)

And the play includes numerous moments when by way of narrative the audience is made aware of the difference between mimesis and reality, as, for example, at the end of Act I: "They all turned around and they saw the ones that looked like Anita Loos Gertrude Atherton Lady Diana Grey Katharine Cornell Daisy Fellowes Mrs. Andrew Greene, and then as they looked the curtain fell not between but so that no one could see any of them. Curtain" (Stein 428).

Stein's narrative voice (or voices) insists throughout on the distinction between characters as these would be represented in conventional theatrical performance, and actors "who looked like Gertrude Atherton Anita Loos and Katherine Cornell." The narrative voice dominates the work so completely, in fact, that in her production of *Not and Now* Tierney decided not to use actors at all to represent the various old-time Hollywood celebrities. Instead, the women were represented merely by a number of empty evening dresses, while the men, in turn, were "present" only as empty suits. Throughout the performance, Tierney and Eileen Green manipulated the dresses and suits according to various prompts in the text, puppet-like, while all the speeches were read by Anne Thulin in something like the manner of a documentary. Whether or not Tierney intended to emphasize the emptiness of the Hollywood lifestyle or the mindless contemporary cult of celebrities, formally the effect was to emphasize "the presence of the absence," which, as Charles Jencks wrote some years ago, "is such a well-known convention of both Late- and Post-Modernism...that it requires no elaboration" (Jencks114).

One final example of the anti-mimetic impulse on the modern stage, this from the work of the contemporary American playwright Ntozake Shange. Her "theater piece" *spell #7* (1979) dramatizes the history of race relations in the United States by reference to the crippling stereotypes—or "spells"—cast on blacks by white society. Shange's metaphor for this act of cultural usurpation is the minstrel show, a form of theatre in which white actors donned "black-face" masks to perform black identities. The most obvious instance of diegetic stagecraft would, of course, be the play's narrator/magician, Lou, who, "dressed in the traditional costume of Mr. Interlocutor," transitions freely between audience and stage (Shange 608). Like Shakespeare's Prospero or like Tom in *The Glass Menagerie,* a character such as this offers mimetic complexities in his own right and stands at some kind of representational crossroads between on and off stage. Lou is as much spectator to the action as its instigator; of particular interest is the way he folds the voices and experiences of unknown others into his own. The following speech, for example, is addressed directly to the audience, but in it Lou mixes a number of different times, points of view, and social contexts:

why don't you go on & integrate a german-american school in st. louis mo. / 1955 / better yet why dont ya go on & be a red niggah in a blk school in 1954 / I got it / try & make one friend at camp in the ozarks in 1957 / crawl thru one a jesse james' caves wit a class of white kids waitin outside to see the whites of yr eyes / why dontcha invade a clique of working class italians trying to be protestant in a jewish community / & come up a spade / be a lil too dark / lips a lil too full / hair entirely to nappy / to be beautiful / be a smart child trying to be dumb / you go meet somebody who wants / always a lil less / be cool when yr body says hot / & more / be a mistake in racial integrity / an error in white folks' most absurd fantasies / be a blk kid in 1954 / who's not blk enuf to lovingly ignore / not beautiful enuf to leave alone / not smart enuf to move outta the way / not bitter enuf to die at an early age / why dontchu c'mon & live my life for me / since the dreams aint enuf / go on & live my life for me / i didnt want certain moments at all / i'd give em to anybody…awright. alec. (Shange 609)

Lou's monologue registers the felt experience of a racist era. Even though his words are formally his own, in their emotional structure they belong also to the various historic characters whose points of view they represent. The structure of the speech blurs these individual experiences and viewpoints with Lou's own more distant stance as interlocutor. The confusion and shame of a black child whose relatively light skin makes him an outsider with respect to both his own race as well as whites, for example, is spoken with the accents of a deeply personal self-realization: "be a blk kid in 1954 / who's not blk enuf to lovingly ignore / not beautiful enuf to leave alone." Likewise the audience is urged to "try & make one friend at camp in the ozarks in 1957 /crawl through one a jesse james' caves wit a class of white kids waitin outside to see the whites of yr eyes." Lou's words have some of the characteristics of choral speech; their temporal simultaneities call to mind a complex social history, somewhat in the manner of the haunting, indeterminate voices of some of Eliot's poems.

But perhaps the most striking instance of diegesis in *spell #7* is visual, "a huge black-face mask," according to Shange's stage directions, "hanging from the ceiling of the theater" (Shange 608). The mask looms overhead as the audience enters the theatre; it is raised out of sight once the play begins, but it appears symbolically between acts and again at the end of the play. Shange's use of the black-face mask as a metaphor for her theatre is, of course, relevant to her particular subject: the black-face mask worn by minstrel performers and also the sterilizing "masks" imposed by white America on a minority population. But Shange's deployment of masks in *spell #7* has a more specifically dramaturgical function. Broadly defined, masks and the performance styles associated with masks have formed an important and influential source for theatrical innovation in the twentieth century. "[T]he mask," writes Susan Smith, "is used so extensively and diversely by so many groups and individuals in different places at different times that its central importance for all of modern drama has not often been understood.… [From 1900 to 1980] fifty dramatists of note and as many directors, designers, and minor dramatists used masks in over two hundred plays and productions, many of which were influential or radical experiments in the theater" (Smith 177). Masks dominate the early, experimental theatres of artists such as Craig and Artaud, just as they figure in Brecht's thinking about "alienated" styles for popular performances. Even when they are used in works that clearly valorize a poetic text as the basis of theatrical performance, such as the plays of Yeats or Jean Cocteau, the masks clarify and focus the meaning of the verbal. And masks are widespread through the drama of the middle and later twentieth century in plays such as Peter Shaffer's *Equus*, in the avant-garde theatre of Europe and America in the 1960s, or in the frozen, impersonal faces that Beckett invariably specifies for performers.

The chief virtue of masks lies in their insistent presence as a signifying vehicle. That is to say, a mask turns any actor who wears it into a sign. Especially for modern audiences used to the conventions of realism in the theatrical or cinematic arts, masks are irreducibly non-mimetic; on stage, as elements of a compositional whole, they can be understood to work like the medieval painting technique known as chrysography, in which actual gold was used to highlight parts of the surfaces of objects in paintings such as leaves or the Virgin's robe. Because the gold tracery physically reflected the ambient light in which the painting was viewed, it stood out as materially different from whatever lighting conditions were depicted in the representation; the reflective surface of the metal brought to the representation an unnaturalness that was impossible to integrate into the pictorial illusion (Hills 26). Masks on stage behave in the same, technically inflexible way. They are either present or not present, and, if present, they cannot by their nature readily be naturalized or "corrected" in the direction of the mimesis because they retain their primary function as a signifying entity.

In this case the "grotesque, larger-than-life misrepresentation of life" (to use Shange's description of the prop) is very much a Brechtian "alienation effect," part of the visible machinery of theatrical representation. Shange's analytical dramaturgy is a reminder of the extent to which Brecht's ideas for a theatre in which mimesis is conditioned by diegesis constitute what Fredric Jameson calls a continuing if "subterranean" influence on postmodernist theatre aesthetics: "[T]he framing of artificial arguments and reasons why Brecht would be good for us today and why we should go back to him in current circumstances seems hypothetical in contrast to the concrete demonstration that we have in fact 'gone back to him' and that his thought is present everywhere today without bearing his name and without our being aware of it" (Jameson 171).

CHAPTER 3
THEATRES OF ABSENCE

Here [India Song], you have the auditorium, you have the stage, and you have another space. It's in this other space that the things are ... lived, and the stage is only an echo chamber.
—Marguerite Duras, interview published in *Woman to Woman* (1987)

One of the most bold attempts to extend theatre to the space offstage occurs in plays that foreclose any attempt whatsoever at direct enactment of people or events; the simplest examples of such dramas are those where one or more characters are thematically central yet remain materially unrepresented. *Waiting for Godot* is certainly the most famous example of such works, its notoriety being such that it is sometimes believed that Beckett's play is sui generis. But plays about characters who never appear on stage are far more common in theatre history than might first be thought. If anything, Beckett's drama is not so much an exercise in novelty as a modern instance of a dramaturgical experiment that has been repeated numerous times throughout Western theatre history. Plays structured around a centralized "absent presence" include mainstream classical comedies such as Terence's *The Girl from Andros*, or *The Pot of Gold* and *Casina* by Plautus. Each of these three works features an important woman character who remains out of sight even as she is an object of active curiosity or gossip on the part of the people on stage.

The deliberate withholding of important characters from sight is common in twentieth-century drama. Susan Glaspell often structures what are essentially realist works around one or more missing persons; plays such as *Bernice* (1919), and *Alison's House* (1931) make use of characters who are entirely figures of spectators' imaginations, as does *Trifles* (1916), her most widely known work. All are plays about women who for both thematic as well as dramaturgical reasons are kept off stage. Fernando García Lorca pursues a similar technique with *The*

House of Bernarda Alba, a play that dramatizes the power of men (or "masculinity") by excluding all males from the stage. Lorca allows for their presence only by means of offstage sounds that represent masculinity or male sexual desire: at one point spectators hear the singing of a group of reapers coming as if from nearby fields, and at another point some of the women on stage are startled by heavy thuds from the hooves of a stallion who is apparently trying to kick down the walls of his stable to get access to the mares in heat. In this way the men who inhabit the imaginary spaces of Lorca's drama, even though they are not seen, constitute nevertheless an important, even central, part of the text: they are required by the play (even though only one of them, Pepe el Romano, is actually given a name), and they can be said in some way definitely to be "in" it.

Absent central characters of this sort can be found throughout more abstract theatre pieces written during the "postmodern" half of the twentieth century, in works such as Marguerite Duras' *Savannah Bay* or Maria Fornes' *Fefu and Her Friends*. In the former, the "play"—like most of Duras's works, it is more a staged recitation than a conventional drama—takes the form of a protracted conversation between two women who attempt to "summon up the memory of the girl who died in the warm sea of Savannah Bay" (Duras, *Four Plays* 112). In *Fefu and Her Friends*, similarly, men are frequently referenced as objects of envy, bewilderment, or amusement, but they never actually appear on stage. In this latter drama, as in *The House of Bernarda Alba*, sexual politics are based on the fact that women have bodies and are, therefore, seen to be tragically vulnerable, while men, being unrepresented visually, are not.

It may be significant that the majority of works that affirm the presence of individual subjectivity by denying the possibility that such subjectivity can be embodied on stage frame questions of sexuality and gender. This principle is true of both ancient as well as more recent dramas. More often than not it seems to be women characters who are concealed from spectators' view, and it may well be the case that the relative invisibility of women in such dramas is related to the larger cultural "invisibility" of women in Western theatre history generally. This would be my own view, though I think the corollary assumption that such staged occlusions are consistent with a history of patriarchal determinism would be too simplistic. Whether the persons who are not given staged embodiments are in any given instance male or female, what is to be stressed is that important and dramaturgically powerful characters in all these plays are not merely hidden but instead are represented in terms of their hiddenness. The status of such characters in a play, in other words, depends less on a particular set of narrated instructions for developing a full mental recomposition or "picturing" of them, as was the case within the general tradition of ecphrasis as it functions on stage as a stimulus to spectators' imaginations, than on the relative, or even absolute, unrepresentability of such characters in the first place. Theatrical power is then born, paradoxically, of the deliberate act of concealment.

* * *

Students who read *Waiting for Godot* for the first time are often surprised by how frequently Beckett references the physical characteristics of somebody who has become famous for never showing up. The play, in fact, contains more than enough data for an imaginative reconstruction of its titular character; whoever or whatever Godot is, he is certainly reclusive, but he doesn't seem to be particularly mysterious. Each of the two acts of the drama includes numerous specific details about Godot, and these details are in the aggregate familiar, even homely, rather than alien or remote. Godot can be confidently assumed to be a man at least middle-aged: his beard—if the memory of the boy in Act II is to be trusted, and there is no reason not to trust it—is white rather than black. Godot lives "in the quiet of his home." He is said to consult his family with regard to his affairs and to maintain a circle of acquaintances: he is described as having friends, agents, and correspondents, and in his daily life he seems to be fully anchored in the material world with his books and his bank account. Whoever he is, this Godot, he has a horse and keeps goats and sheep; his household, therefore, presumably includes some amount of rural land and outbuildings. There is a loft, there is hay. Even without Beckett's specifying an exact time and place for the action, it is possible to fill in with a relatively high degree of realism the picture of Godot that emerges from this assemblage of biographical data. It is the picture of a moderately successful entrepreneur living somewhere in western Europe in modern (or relatively recent) times, a gentleman farmer or perhaps a squire.

The objects imagined to be associated with Godot—home and family, friends, bank accounts, and business dealings—are all centered on middle-class domesticity. Initially, neither Godot nor the situation that Didi and Gogo find themselves in seems particularly unusual or sinister. These are not the behavioral characteristics of an autocrat or despot, much less an Übermensch or deity. As for Godot's failure to keep his scheduled appointments, this also is part of the fabric of modern living, not necessarily ominous or unrealistic; one encounters much the same kind of frustration dealing with plumbers and roofers. There is in the general description of Godot and his habits one slightly ominous note (ominous for contemporary readers, at least): in a description whose language seems to reach back to a bygone age where corporal punishment was understood to be a necessary part of civilizing children ("spare the rod and spoil the child"), Godot seems to have the freedom to be arbitrarily violent. The boy at the end of Act I says he keeps Godot's goats, and he claims that his master beats his brother, who keeps the sheep, but not him:

> *Vladimir*: You work for Mr. Godot?
>   *Boy*: Yes Sir.
> *Vladimir*: What do you do?
>   *Boy*: I mind the goats, Sir.
> *Vladimir*: Is he good to you?
>   *Boy*: Yes Sir.
> *Vladimir*: He doesn't beat you?
>   *Boy*: No Sir, not me.
> *Vladimir*: Whom does he beat?
>   *Boy*: He beats my brother, Sir.
> *Vladimir*: Ah, you have a brother?
>   *Boy*: Yes Sir.
> *Vladimir*: What does he do?
>   *Boy*: He minds the sheep, Sir.
> *Vladimir*: And why doesn't he beat you?
>   *Boy*: I don't know, Sir.
> *Vladimir*: He must be fond of you.
>   *Boy*: I don't know, Sir. (Beckett 55–56)

The first thing to be said about the boy's recollection of his encounters with Godot is that it is itself a primitive messenger's speech. Even though it is broken into pieces so as to be rendered as a series of questions and responses rather than in the form of a conventional narrative, it is still a translation of first-hand experience and witnessing of offstage events into the language of story. And like any story, therefore, it "re-counts" previous events into minimally segmented units that are presented to listeners (both on- and off-stage) as having taken place at a specific point or over a specific temporal interval (Fleischman 101).

One important pragmatic difference between the boy's discourse and a more conventional messenger narrative is that it lacks the characteristic orientation of narrative toward exposition; it is more like an objective report than a coherent, interpretive story. As Fleischman puts it, "listeners [to a story] should not be left wondering 'what's the point?'" (Fleischman 103). But whether report or story, the overall purpose of the boy's commentary is still to provide a cognitive frame for what he has himself personally experienced. The boy comes on stage to assure the audience that despite his remoteness Godot is real, although he exists somewhere beyond the margins of what can be brought within view. The distinction between the boy who comes on stage and his off-stage brother is given, significantly, a biblical frame: whoever or whatever Godot might be, he hands out punishment based on the millennial vision of Christ culling sinners from those whose souls are saved, just as a shepherd would separate out the goats from his flock of sheep. It is, incidentally, characteristic of Beckett's noncommittal attitude with respect to religion (he surely had the well-known passage in Matthew in mind as he composed this scene) that he would reverse the biblical value system in his play so as to make the keeper of goats appear to be the more fortunate of the two herdsmen.

In this and numerous other passages where the talk turns to the habits and whereabouts of Godot, Beckett's play contrasts the imagined potency of a character whose place is out of sight with the comparative powerlessness of those characters who can be seen. Godot's existence somewhere offstage is played off against events in the lives of characters who are materially present on stage and who speak and act. Mimesis and diegesis run side by side, as it were, functioning as two interdependent modes of theatrical representation. By their mimesis, Vladimir and Estragon (as well as the other pair of mimetic characters, Pozzo and Lucky) affirm repeatedly the painful realness of their own existence: we are, we are here. Their final instruction to the boy on his second visit, at the end of the play, is to have him confirm to Godot their visibility:

> *Boy*: What am I to tell Mr. Godot, Sir?
> *Vladimir*: Tell him... *(he hesitates)*... tell him you saw me and
>   that... *(he hesitates)* .. that you saw me. (Beckett 106)

It is no accident that the main characters, Vladimir and Estragon, are almost always in view. Vladimir leaves for brief periods only twice (both times to urinate), both in the first act, and Estragon never. The space they occupy, their openness to view, the articles of clothing that they fuss with, their incessant, babbling conversation, all reassert not only their own powerlessness but also the corollary potency of the unseen Godot. This inability to remove themselves from mimesis, to escape the condition of embodiment, causes them considerable anguish. Estragon cannot hide even for a moment; at one point he "*goes and crouches behind the tree, realizes he is not hidden, comes out from behind the tree*" (Beckett 83–84). Their constant exposure mimics the vulnerable condition of Adam and Eve in the Garden once they have eaten from the tree of good and evil. As was true of the relationship between the God of the Old Testament and the first two humans, the basis of Vladimir's and Estragon's relationship with Godot requires an absolute inequality between what can and cannot be made visible.

To be visible—to be embodied—in the context of *Waiting for Godot* is to be capable of being hurt. It is as if physical vulnerableness forms the grounds on which Beckett defines theatrical character. Scenes of pain and physical suffering occur frequently throughout the play, and the implements that actually inflict that pain on the characters—the boot, the whip, the rope—themselves have considerable iconographic status as weapons of torture. It is true that scenes such as these are partly the result of the comic matrix of Beckett's drama, a play that is often called a tragicomedy; pratfalls, violent beatings, and physical deformities are the stock in trade of comic dramatists from Aristophanes to Stoppard. Typically these conventions of the genre are explained in terms of their value as culturally licensed forms of taboo-smashing that carry out useful cathartic experiences for spectators. The theory is that the violence in comedies provides opportunities for audiences to blow off steam, so to speak; the notion of a comic catharsis has proved especially influential among twentieth-century writers on the subject of comedy, even though there is no evidence whatsoever in support of it and some that contradicts it (Konečni 230).

Whether or not watching one person inflicting pain on another can somehow purge spectators of the impulse to do something similar in real life, the typical topoi of comic theatre seem part of Beckett's larger dialectic between embodiedness and vulnerability, on the one hand, and, on the other, invisibility and potency. Quite apart from any accidental or intentional play on the similarities between "Godot" and "God" (a linkage that is, of course, not available in the French in which the original manuscript, *En attendant Godot*, was written), this is the reason that Beckett's play is so often understood from within a religious or allegorical framework. It may seem theatrically naive to say that a character who neither appears nor speaks is "represented" with far greater authority than those who act and talk on stage. But that empowerment and visibility in the theatre should exist in inverse relationship is perfectly consistent with long-standing practices of representing power, both in the visual and the literary arts as well as in Judeo-Christian scriptural tradition that identifies God and His omnipotence as beyond material embodiment. "To have a body," says Elaine Scarry, "is to be describable, creatable, alterable, and woundable. To have no body, to have only a voice, is to be none of these things" (Scarry, *The Body in Pain* 206).

A lesser degree of this principle of non-representability can be found commonly in Western literary history, where major characters sometimes acquire a measure of authority by remaining out of sight in order to be made the subject of other characters' speculations or narratives. In this case, to be unrepresented is to become an object of imaginative belief. This is why in *Moby Dick*, initially at least, Ahab acquires a larger-than-life status because he is a rumor rather than a flesh-and-blood event (he does not make his appearance until the Pequod is several days at sea). It is why, also, in Shakespeare's tragedies one first hears gossip about Lear, or Macbeth, or Antony and Cleopatra before they step into view on stage. But the most extreme demonstration of what Scarry calls "the error of representation" (Scarry, *The Body in Pain* 227) are the various scriptural accounts of the God of the Old Testament, where, for example, the many narratives about Jaweh (itself a non-word, a word, that is, chosen to "represent" the deity precisely because it is mimetic of nothing) stress that He is in all ways beyond representation. Even in those passages where God manifests His presence by taking on some material form—a flame in a bush, a whirlwind—the most important characteristics of God are still his hiddenness and his disembodiment.

That the representation of characters or events should be made dependent exclusively on their being made the subject of others' imaginations, that it can be a theatrically efficacious strategy to dramatize a person or an event by choosing deliberately *not* to represent it materially, nor even directly to describe it, is the method pursued by the playwrights I discuss in this chapter. On these stages it is as if (as was suggested once by Jacques Derrida) "pure absence—not the absence of this or that, but the absence of everything in which all presence is announced—can *inspire*..." (Derrida 8).

* * *

The comedies of the Roman playwright Plautus now and then include characters who are important to the plot but who never appear on stage. Typically these characters are women. There is one such off-stage character in *The Pot of Gold*—Phaedria, the daughter of the miser Euclio. *Casina*, Plautus' adaptation of a Greek play by Diphilus called *Kleroumenoi* ("The Lot-drawers") also centers on a woman who remains off stage. The characters and events in both *Casina* and *The Pot of Gold* are similar; "the plot structure of Greek New Comedy," Northrop Frye once wrote, "as transmitted by Plautus and Terence, is itself less a form than a formula" (Frye 163). *Casina* was first staged between 186 and 184 BCE, the year of Plautus' death; the play develops the story of Casina, a young woman thought to be a slave, though she is actually a free-born Athenian. In *Casina*, there are actually two characters who are important to the plot but who never appear on stage—in addition to Casina, there is also a young man, the son of Lysidamus, who (it is reported) plans to marry her once it is discovered that she is a free-born citizen of Athens.

The play as a whole seems conceived as an amusing variation on its New Comedy antecedents. It takes the familiar competition between an old man and a young one for the same woman and recasts it as a contest between the respective slaves of Lysidamus and his son for a house-hold slave girl; by focusing his play on the love affairs of slaves, Plautus in effect alienates the comic Oedipal drama that so often drives the plot of New Comedy. Other features of the text indicate that in adapting the play for Roman audiences, Plautus intended to have even more fun with the stock materials of the genre. The speaker of the Prologue, for example, announces to his audience that a person one might rightly expect to have an important place in the drama is not even going to appear on stage; Lysidamus has sent his son abroad, and, we are told, "[h]e will not return to the city today—do not expect him—during the course of this comedy. Plautus would not have it so—he broke down a bridge that lay on the youth's route" (Plautus 9). This humorous, self-reflexive gesture on the part of the author suggests that the problem of crafting a play with characters who do not appear on stage may have provided an interesting creative challenge to the playwright. Plautus does not even bother to give the young man a name (it is Euthynicus) until the Epilogue, even though commentators generally agree that the Greek original included scenes involving Euthynicus' homecoming (his name means "victorious") as well as his marriage to Casina.

There are several scenes in the play where Plautus structures the dialogue so as to bring virtual images of the absent Casina before spectators. At the end of the first act, for example, Olympio, one of the slaves in competition for Casina, taunts his rival Chalinus by telling him a sexual fantasy that he frames in part from the point of view of the off-stage woman:

> You shall be fastened tight in the window-frame where you can listen while I'm kissing my Casina. And when she says to me: (*in languishing accents*) "Oh you little darling, my Olympio, dearie, my life, my little honey boy, joy of my soul, let me kiss and kiss those sweet eyes of yours, precious! Do, do let me love you, my day of delight, my little sparrow, my dove, my rabbit!"—when she is saying these soft things to me, then you'll wriggle, you hangdog, you, wriggle like a mouse, in the middle of the wall there." (Plautus 17)

Another narrative sequence involving Casina occurs near the end of the third act, when Pardalisca, the maid of Cleostrata, Lysidamus' wife, describes the recent behavior of Casina when she learns that she is to marry someone she doesn't want to:

Lys: What's inside? What is it?
Par: She's following the wicked manners of wicked women and threatening her own husband. It's his life—
Lys: (*alarmed*): Well, what, what?
Par: Ah-h!
Lys: What is it?
Par: —it's his life she wants to take, so she says. There she is, a sword–
Lys: Whew!
Par: —a sword—
Lys: What about this sword?
Par: —in her hand!
Lys: Lord preserve us! What has she got that for?
Par: She's chasing everyone through the house there, and won't let a soul come near her; they're hiding under chests and couches afraid to breathe a word.... But oh, sir, if you only knew what she said this day—
Lys: That's what I'm anxious to know. What did she say?

Par. Listen, sir. She swore by all the powers above that she would murder the man she spent this night with. (Plautus 71–73)

Finally, even though there are some passages missing from near the end of the manuscript, it is clear that the last act of *Casina* features a hilarious parody of a tragic "messenger speech" in which Olympio recounts his attempt to have sex with Chalinus who is disguised as Casina:

> Attention, now, while I give you an account of myself; it is worth your while to lend your ears. Oh, it's comical to hear of, and to tell of–the mess I made of things in there! When I led this bride of mine inside I took her straight off to a chamber. But it was dark as a dungeon.... Then I call her by name: "Now, now, Casina," says I, my own little wifey, what makes you so cruel to me, your own husband?...When I see she's barricaded herself, I beg her not to be so awfully coy....A nice long kiss * * * and I get my lips punctured by a beard that's just like bristles, and the next instant, as I'm kneeling beside her, she rams both feet through my chest. (Plautus 97–99)

Terence's story of the woman from Andros provides another instance where dramatic presence depends on absence; his experiment with an off-stage character is somewhat bolder than those of Plautus. *Andria* (166 BCE; translated either as *The Woman from Andros* or *The Girl from Andros*) is generally thought to be Terence's first play, produced when he was only nineteen. As was the custom among Roman playwrights, Terence's work was spun off a Greek text, in this case two of Menander's plays with almost identical plots. The plot centers on a romance between a young man named Pamphilus and Glycerium, a young woman whom everybody believes to be the sister of Chrysis, a local courtesan. Chrysis and Glycerium have been living for the last several years in Athens, but both women came originally from the island of Andros. Chrysis has died before the play begins, and Glycerium is pregnant by Pamphilus, who has promised to marry her. But Pamphilus' father, Simo, has already contracted for him to marry Philumena, the daughter of his rich friend Chremes. Simo suspects, however, that Pamphilus prefers Glycerium to Philumena (who herself happens to be in love with another young man, Pamphilus' friend Charinus). But Simo resolves to press forward with the wedding plans in order to test the genuineness of his son's affections. Meanwhile Davus, Simo's household slave, plots secretly to help Pamphilus thwart his father's plan for his marriage. When it becomes known to all that Glycerium has given birth to Pamphilus' child, Chremes angrily breaks off the match between Pamphilus and Philumena, and everybody's schemes come to a standstill. Just when all hope seems lost, however, Chrysis' cousin, an old man from Andros named Crito, arrives with the news that Glycerium is actually Chremes' own lost daughter. Because of her new-found wealth and social status, Glycerium is now free to marry Pamphilus, and the happy Chremes gives his other daughter Philumena permission to wed Charinus.

Terence's play has had two well-known reprises. The first, Niccolo Machiavelli's translation, *The Woman of Andros* (1517), closely follows the dramaturgy of the original. The second is a much freer version by Thornton Wilder (*The Woman of Andros*, 1930) who transformed the comedy into a serious, meditative work of fiction about love and loss at the dawn of the Christian era. Centuries of commentary on Terence's play focus typically on the ways in which his work diverges from its Greek antecedents as well as on the playwright's deftness in managing the subtleties of character and intrigue that were the stock-in-trade of Greek New Comedy. But among the more interesting aspects of the commentary on Terence's play is a difference of opinion as to which of two women from Andros, Chrysis or Glycerium, was meant to be referenced in calling the play *Andria*, "the woman from Andros." Most scholars since the early modern period have assumed the woman of the title to be Glycerium. It is true that Chrysis is described a few times early in the play as "the woman from Andros" (Terence 13). But she is almost never mentioned after the exposition in the first act, and it seems more likely that in devising a plot for his comedy Terence would have been thinking of a living, younger woman as the ongoing object of so many men's desires and schemes. Glycerium is the younger and more beautiful of the two women; she's the woman Pamphilus impregnates, she's the one he (and other characters) love or lust after, and she's the one who turns out to be the Cinderella, so to speak, who marries the prince. Besides, when the play opens, Chrysis is already dead; to argue that she is the main character would be like saying that Shakespeare centered his tragedy on Hamlet the father, not the son. But a small and eloquent minority (and Thornton Wilder is among them) have argued that it is Chrysis who is the woman from Andros. Wilder once remarked that "I have always assumed that the title applied to Chrysis" (Haber 35), and in restructuring the classical drama into a narrative, he made Chrysis the focus of the action, representing her as living for much of the novel. Even after her death, she remains the dominant character.

In recasting Terence's drama as a novel, especially in making Chrysis into the most important character, Wilder was able to take advantage in particular of the double time structure intrinsic to all narrative: the time the story is told, on the one hand, and, on the other, the time when the events of the story are understood to have taken place. Like most novelists, Wilder makes use of each of these two separate times (sometimes called, respectively, "speaker-now" and "story-now") in creating a prospective orientation toward characters and events that belong technically to the past. To the extent that the third-person narrator in Wilder's novel often gives the impression of a story that moves from earlier events to later ones, one can readily believe that Chrysis is alive and "moving forward" in time. Another (and more formal) way to put this would be to say that a distinctive feature of narrative—one not easily transferrable to drama— "is its ability to recreate the experience of the events, in other words, to replicate *post hoc* the contingent prospection of the current report form—whose data source is not memory but direct perception" (Fleischman 131). In narrative, for example, widely separated personal or historic times can easily be juxtaposed or made to overlap in ways that are difficult if not impossible to achieve on stage, and the ease with which the narrative work of literary art may be temporally "stratified" (to borrow Roman Ingarden's term) in this way is one of the genre's chief distinguishing features.

In practical terms, therefore, narrative is much better able than drama to represent characters who are no longer alive at the time when the central action takes place. In a play, the dead are conventionally given embodiment only as ghosts (if they are to be embodied at all) or referenced retrospectively in speech. But a narrator may easily bring the dead to life, so to speak, simply by modifying the temporal framework of his or her story, switching from the fundamentally retrospective point of view of narrative to one where the same events are seen as if from a prospective orientation, as in the following two narrative variations of the same sequence of actions (cited by Fleischmann); it will be seen that the second version of the narrative contains two different temporal orientations on events, one conventionally retrospective and a second, "prospective" point of view:

(1) President Reagan *arrived back in Washington*...at noon on Monday. Four hours later he *flew to Camp David*...where he *met with National Security advisors.*

(2) President Reagan arrived back in Washington at noon on Monday. *In four hours* he flew to Camp David, where he met with National Security advisors. (Fleischman 129)

Terence's comedy, like *Waiting for Godot*, is structured as an exercise in the sustained imagining of a person who does not appear on stage. As to whether an audience is supposed to picture Chrysis or Glycerium somewhere in the offstage, the following commentary is based on the assumption that Glycerium, not Chrysis, is the imaginative and thematic center of the comedy. But whether one takes Glycerium or Crysis to be the Andrian referenced in the title makes little difference from a compositional point of view. Indeed, it may be that Terence was aware that there were in his play two "women of Andros" who would remain unseen. In any case, the playwright is faced with the task of installing at the center of the drama a character (or characters) who for calculated thematic or dramaturgical reasons does not—or, in the case of Chrysis, cannot, unless she is represented as a ghost—come on stage.

Like Godot, "the woman from Andros" is to be conceived as existing in a determinate (if largely unspecified) space "offstage." But the playwrights' reliance on language rather than enactment in both plays ensures that these missing persons will remain materially unrepresented and unrepresentable. All the events involving Glycerium—her childhood, her life with the courtesan Chrysis, her love affair with Pamphilus, even her discovery of her true identity—take place in the offstage, either before the beginning of the represented action or synchronous with it. In this way, paradoxically, she exercises considerably more hold on the imagination than if she were to appear. Just as it would be relatively easy for a living actor to match the various descriptions of Godot and his habits, the subject here—a young woman—would be easily representable according to the theatrical conventions of Roman stage comedy. Yet for some reason Terence consciously excludes her from the stage. In both cases it is precisely the characters' absence that makes him or her a rich object of speculative thought.

A linguistic construct that exists solely in the domain of the onstage characters' fantasies has none of a living actor's restraints and can be made to conform to whatever beliefs about women or feminine attractiveness spectators might harbor. In a situation like this, where "seeing" in the theatre is dependent specifically on absence, the character or event spectators are asked to imagine almost always takes shape based on recollections of a category of similar experiences—experiences that, of course, vary widely from one person to another. This is different from ecphrasis in which the spectator is given a relatively full set of instructions with which to imagine offstage persons or events. Conventional messenger speeches tell spectators how to see an offstage event and in most cases what to make of it, but in Terence's play they are obliged to construct a picture of the woman from Andros without anything like the elaborate verbal progressions involved in Euripides' narrators' accounts of events. In cases where fuller descriptions are lacking, writes Gilbert, the brain acts "like a portrait artist commissioned to produce a full-color oil from a rough charcoal sketch, filling in all the details that were absent" (Gilbert 99).

It is important, however, that spectators' imaginations about the woman from Andros not be left entirely uninstructed; thus the imaginability of Glycerium in part rests on the frequency with which other characters speak of her. Even though Glycerium never steps into view, she, like Godot, is present often on the lips of the rest of the characters. The result is that she is simultaneously everywhere and nowhere, and over the course of the play she acquires, ultimately, a rich personal history. She is described initially as one who is "demure and charming, quite beyond compare" (Terence 15), and she is placed within a family context as Chrysis' sister. At different times and in different contexts she is described as one who walks dangerously near a funeral pyre (Terence 15); she is heard to cry out to Juno during labor (Terence 36); she is one who as a little girl was shipwrecked with her uncle on Andros (Terence 65); and one who was originally named Pasibula, the daughter of Chremes (Terence 67). Indeed this absent woman is referenced so frequently over the course of the drama that she acquires something like the status of a legend. Even though she is not ever seen, the different stories that are told about her verify the material presence of a human body existing through time. As a result, even without any actual sight of her, or even without any graphic description to assist an audience in constructing a sensory picture of her, audiences can nevertheless generate an idea of her living in the extrascenic space just behind the doorway of her house. She becomes less a character, as that term is normally understood, and more like an object of belief—a central element of Terence's dramaturgy, despite the fact that no actor stands in for her on stage.

These wholesale displacements of female characters to the offstage occur in the drama of a society widely known for its political and economic subjugation of women, and it is worthwhile to dwell for a moment on the question of the extent to which plays such as *Casina* and *The Girl from Andros* are representative of a dismissive ignorance with respect to the lives of women in Roman society, perhaps even a diffuse misogyny. Like Plautus, Terence wrote for a theatre where women were literally invisible—their parts were played customarily by male actors—and for a culture where women had few real political or property rights. It's tempting, therefore, to suppose that Roman stage practices in the second century BCE, working in conjunction with a pervasive belief in the second-class status of women, could certainly have given rise to this kind of "celibate" mode of theatrical representation or at least been consistent with it. If you believe that real political identity always entails the rights to be seen and to speak for oneself, Terence's comedy indeed looks singularly misogynist.

But could it be possible that the dramaturgy in Terence's play might cause an audience to take a *less* restrictive view of women and their place in society? The practice of having men play women (or women, men—as in the case of Hrotsvitha's dramas) is not necessarily always indicative of repressive sexual politics. Cross-dressing is widespread throughout theatre history. It appears in a great variety of social and performative contexts, and many of the world's great theatres, in Asia as well as in the West, have been based on it. The custom appears so frequently in theatre, in fact, that if alien eyes were to look at world theatre history, cross-dressing might appear not at all irregular or curious but something more like a default position. Of course, there is no making theatre without politics, even (perhaps especially) sexual politics; but it is theatrically naive to assume that having men play women (or women men, for that matter) is always simply and uniformly exclusive with respect to the sex that is being enacted. The widespread belief that an actor's sex or ethnic type must match that of the character he or she plays derives from modern preferences for naturalist performance styles (a preference influenced no doubt by a century-long history of exposure to photography and to cinema) as well as from simplistic assumptions about human identity.

In the case of the classical stage, the contemporary conviction among some critics that no male could legitimately represent a sexually mature woman (or a woman, a man) rests on the relatively modern belief that in representing "character" the actor brings into view a unique, historic, inviolate self. However, to the extent that identity (or gender) is understood in some ways to be a mode of performance, subjectivity becomes more ambiguous, more expressly an act of theatre, and it is this broader, more flexible view of selfhood that makes it theatrically effective for one sex to represent (or to "pass for") another. Indeed, if having men take over the parts of women is a kind of theft, it can—as Eric Lott has shown with respect to the complicated and ambiguous history of "black-face" or minstrel performances—be also construed as a gesture of intimacy, even love (Lott 3–12). Certainly where "character" is understood in broader or more flexible terms, as is often the case, for example, in Asian theatre traditions, the glaring lack of "fit" between the gender of an actor and that of his or her character tends not to be occlusive or unfairly expropriatory. On the contrary, it is a singularly theatrical gesture, and for this reason it can just as well remind spectators of the mimetic or performative acts that produce identity in the first place.

It helps, therefore, to see Terence's moving of the women to the offstage as a dramaturgical strategy in some ways similar to Beckett's. Both plays depend on a decisive and fully conscious turn away from mimesis as the ground of theatrical representation. Even though Glycerium is never actually seen, there is one scene in which she is heard to cry out, in childbirth. It comes just after the beginning of the third act. Simo and his slave Davus, stand on the street just outside Glycerium's house; independently, they lay bare their thoughts and their schemes to the members of the audience. It begins as a scene in no way different from a multitude of others in New Comedy, where scores of clever slaves (the theatrical descendants of the prime mover of ancient comedic forms, the *eiron*) weave fantasies of power and intrigue in open disregard of the authority of their masters. The dialogue proceeds by means of a sequence of asides designed to expose Davus' schemes to the audience:

> *Simo* (Aside): How can it be? Is he out of his mind?
> A child by a foreign woman. Oh, wait...
> Yes, that's it. I've just caught on, at last.
> I was stupid.
> *Davus* (Aside): What's he say he just caught on to?
> *Simo* (Aside): Here we have Davus' first clever maneuver:
> They pretend she's giving birth, to scare off Chremes.
> (Terence 36)

For those Romans who had assembled before the temporary stage that would have been erected for the performance of this first comedy of Terence, the interaction of these two stock characters would have sounded both intriguing and yet completely familiar; it must have been a scene as delightfully predictable in its comic promise as, for modern audiences, the moment when Bugs Bunny first spots Elmer Fudd decked out in his hunting clothes, strolling through the meadow. But the conversation is cut short by a cry that comes from beyond the limits of vision:

> *Glycerium* (From inside the house): Juno Lucina! Save me. Help, I pray.
> *Simo* (Aside): So sudden, eh? After she heard that I was here in front of her door, she got busy. (*To* Davus) Davus, your timing's way off. (Terence 36)

It helps to imagine this sequence as it might be approached by actors who were concerned to solve problems of timing and affect. The scene makes no sense unless the actors playing Simo and Davus visibly interrupt their dialogue to register a reaction to the shouts coming from within the house; for example, in the performance at the Ludi Megalenses in 166 BCE, suppose that the actor assigned to play the part of Simo has reached that portion of the text where he is about to accept the deception that the story of Glycerium's pregnancy is a hoax: "How can it be? Is he out of his mind?" The actor pretends to mull over the situation; he speculates out loud and then, in response to an apparent revelation, assures the audience that at last he has seen through the trickery. Perhaps he even leans toward the assembled spectators, bringing his hand to his mouth to complete the illusion that he speaks to them in strict confidence: "I've just caught on, at last. I was stupid." During this entire sequence of speeches the actor's body has been fully consonant with the words he speaks; what the character says is to a significant degree authorized by the body of the man who represents him.

Suddenly the actor playing Simo must take into account sounds coming from off the stage. Even though he has no doubt been anticipating hearing them—one must assume the scene to have been repeated at least a few times during rehearsals—his body itself cannot fail to register the impact of the cry. For spectators the jolt of recognition is even more powerful. From their point of view, the hazy prospect of a woman who has previously been denied embodiment now acquires the most deeply incontrovertible proof of actual physical being: an expression of great pain. Onto the imagined body of the woman is projected the immediate sensation of actual, raw sound, so that in the minds of the audience a figure known to them only in their imaginations as "Glycerium" suddenly comes to possess something like full weight and solidity. Because the woman's body is wholly absent and her identity is expressed only as a cry of agony heard from behind a closed door, her invisibility, experienced in conjunction with her sudden eruption into speech, becomes metonymic of her entire existence. It is a radical and concussive theatrical gesture, wholly dependent on the imagination: no actor visible on stage, neither male nor female, could convey the full burden of Glycerium's desperation. In much the same way, on the modern stage, no sound could be as expressive of human agony as the awful "silent scream" voiced by Helene Weigel in Brecht's *Mother Courage*, when she depicts the tragedy of a mother who cannot grieve publicly for her dead son. This particular moment in Terence's comedy takes its place, I would argue, among the several astonishingly "sympathetic" images of women in ancient drama along with Medea's defiantly political assertion that giving birth to a child is more terrible an experience than marching three times into armed combat.

To assume that removing a female character from sight (it is immaterial in this case whether one assumes Glycerium or Chrysis to be "the woman of Andros") is only a way to deprive her of identity or subjectivity, in other words, is to misread the way Terence's inventiveness makes the offstage seem part of the entire signifying space of the theatre. Only if one takes the odd position that it is somehow easier for a dramatist to write a play about characters who never appear on stage, than to present those characters (in Aristotle's phrase) "as living and moving before us," is it possible to believe that because they are not seen women are somehow supposed to be read out of Terence's comedy. Terence's dramaturgy of missing women should be judged, I would suggest, in relation to the considerable versatility of the diegetic space that was available to him and his contemporaries. Of course, to move Glycerium offstage can be seen as a specific editorial act, whether political or simply dramaturgically expedient; my point, however, is that Terence's conception of the offstage can also be understood to be part of a productive, positive artistic strategy. That is to say, Terence makes use of the offstage not just in the sense of a real space that is to be imagined as adjoining the visible space, as would be true, for example, of the space spectators imagine *Agamemnon* to be located in when they hear the king cry out as he is being murdered behind the closed doors of the palace. Terence's offstage also functions in a more artistically productive capacity; it offers spectators a space for mental stretching in the sense of Gillian Rose's description of feminist geography as a narrative space that "straddles the spaces of representation and unrepresentability" and can, therefore, "acknowledge the possibility of radical difference" (Rose 154).

* * *

In a number of recent discussions of political and social power as they are made manifest in the theatrical performances of ancient Greece and Rome, it is frequently assumed that those characters who are represented on stage and endowed with the opportunity to speak are thereby "empowered," while those who do not speak, or those who are given limited exposure or are excluded entirely from the stage picture, are assumed as a consequence of their absence or silence to lack power or in some way to be disenfranchised. The *locus classicus* of such criticism of ancient Greek drama, for example, would be Sue-Ellen Case's well-known essay published originally in 1985 in *Theatre Journal*: "Classic Drag: The Greek Creation of Female Parts." In that essay, Case writes that "the images of women in these plays represent a fiction of women constructed by the patriarchy" (Case 318). Because the plays in turn were written invariably by men and because the roles of women were played invariably by men, Case sees that women are doubly excluded from them: "The feminist reader might conclude," she argues, "that women need not relate to these roles or even attempt to identify with them. Moreover, the feminist historian might conclude that these roles contain no information about the experience of real women in the classical world" (Case 324).

Initially this view seems to make perfect sense: what, indeed, could be more consistent with the political repression of a person, a group, or an entire class of people than to deny them the opportunity to be seen and heard on stage? It seems contrary to everyday experience, therefore, to assert that a literary *Figur* could solidify his or her hold on the imagination to the extent that he or she does not appear. But as we have seen, this is precisely what happens in *Waiting for Godot, Casina,* and *The Girl from Andros*; in unpacking the dramaturgical method of these plays I have attempted to describe in each of them an intimate relation between plays based conventionally on enactment and those dependent on more narratological structures, between characters who are visible and those who are not. To this point my interest has been largely to show the way the offstage can be used in service of the emotions or imagination; in much the same way that absence is said to make the heart grow fonder, dramas such as these are built entirely on the particular imaginative pleasures that attend the mental activities associated with the brain's frontal lobe (Gilbert 18–21). While my readings of *Casina* or *The Girl from Andros* do not necessarily contradict the broadly "patriarchal" readings of ancient drama such as those proposed by Case and others, I would nevertheless want to attribute to these plays and their audiences a greater degree of dramaturgical subtlety than readers such as Case grant them, and, more important, I would also want to insist on their potential for affective and intellectual expansiveness that more narrowly political interpretations have denied them. That removing entirely one sex of characters from the stage can be a strategy to cause such characters to become *more*, not less, important in the eyes of the audience is a method pursued by a number of modern playwrights whose ambitions for their work are broadly political and sometimes expressly feminist.

Susan Glaspell is one such twentieth-century playwright whose specific interests in dramatizing women's stories also led her to experiment with offstage characters and spaces. Glaspell often made use of absent central characters, to the extent that they become a signatory feature of her work; *Trifles* (1916), her best-known play, includes two such characters who are talked about but never seen, a husband and wife named John Wright and Minnie Foster Wright. The play opens one day after John Wright has been discovered dead in his bed, a rope around his neck, and his wife has been arrested on suspicion of murder and taken to the local jail. There is no real doubt that Minnie Wright strangled her husband; the search instead is for her motive in killing him. The day following the murder, three men—the sheriff, the county attorney, and Lewis Hale, a neighboring farmer who discovered the body—visit the Wrights' house to search for clues and to interrogate two of Minnie Wright's friends, identified as Mrs. Hale and Mrs. Peters. The men are completely baffled as to the reason for the crime; they are insensitive to the household environment and its stultifying atmosphere, but the women, in obvious contrast, are keenly aware of Minnie's desperate loneliness. By paying attention to what they know of Minnie and her husband's repressive treatment of her as well as to bits of evidence that the men dismiss as "trifles," the two women unlock the mystery of why Minnie killed her husband in the particularly cruel way that she did. A melodramatic epiphany comes when they discover Minnie's pet songbird with its neck wrung:

> *Mrs. Peters*: Somebody—wrung—its—neck. (*Their eyes meet. A look of growing comprehension, of horror. Steps are heard outside. Mrs. Hale slips box under quilt pieces, and sinks into her chair.* (Glaspell 917)

Trifles is basically a *pièce à thèse* on women's second-class status and on the differences between the way men and women see and understand the world. Accordingly, the invisible Minnie Wright ("kind of like a bird herself," says Mrs. Hale) has been read as a symbol for all the women who are rendered "invisible" in a society run by and for males. Though Minnie is surely guilty of murdering John Wright, the crime is framed more as her violent but understandable retaliation for the way an insensitive world in a sense "killed" everything that was lively and beautiful in a lonely woman. "I wish you'd seen Minnie Foster when she wore a white dress with blue ribbons and stood up there in the choir and sang," says Mrs. Hale to her friend. "Oh, I *wish* I'd come over here once in a while! That was a crime! That was a crime! Who's going to punish that?" (Glaspell 917).

Clearly, Glaspell wrote her play to comment on women's marginal status in society; but I'd add also that it's important to see a subtler connection between Glaspell's dramaturgy and her politics. Glaspell depicts a world where things that are trivial are understood properly to be critical, and where what is conventionally central becomes marginalized. This inversion makes the act of erasure itself into a positive representational strategy, in that, to the extent that Minnie is denied the power to stand in for herself, that absence becomes the basis for our imaginative absorption with her. That is to say, in *not* seeing Minnie Wright, spectators are likely not only to draw conclusions about Woman's relative political invisibility but also to make of Minnie's non-appearance a space for their own empathic engagement with her.

Much of the dialogue can be read as a set of instructions for constructing an idea of Mrs. Wright. First, these are accomplished through narratives of personal experience, as Hale describes conversations that took place the day before when he came to the Wrights' house in search of John: "'Can't I see John?' 'No,' she says, kind o' dull like" (Glaspell 913). Subsequent conversations between the women summon up images of Minnie through references to her habits ("She worried about that [her jars of fruit preserves]," says Mrs. Peters, "when it turned so cold") or by the metonymy of a rocking chair set in motion (Glaspell 913, 914). In all these episodes, the conscious withholding of Minnie Wright from sight means that the audience will be inspired to shift the site of mimesis from the stage to a cognitive act. Like Terence, Glaspell doesn't really "represent" her central character, she causes spectators themselves to do it.

As further examples of twentieth-century dramas that rely for their chief emotional and thematic effects on the felt presence of characters who are not seen, consider, respectively, Federico García Lorca's *The House of Bernarda Alba* (1936) and Maria Fornes' *Fefu and Her Friends*. *The House of Bernarda Alba* is Lorca's last play, written just a few months before he was assassinated by Nationalist soldiers. Lorca's play, "a drama of women in the villages of Spain," documents several days in the life of three generations of women living in the same house: Bernarda, age sixty; her mother Maria Josefa, age eighty; and Bernarda's five daughters (Angustias, Magdalena, Amelia, Martirio, and Adela) whose ages range from twenty to thirty-nine. There are sixteen separate female characters in the play, counting those identified only generically as "women mourners," but though men are often talked about and sometimes even heard talking and singing in the offstage, they are never actually seen.

The play begins upon the women's return from the morning's funerary rites for Bernarda's husband. In the opening scene of the drama there is depicted not only a household shut up in mourning but also a barren society, as Bernarda evidently does not care that in carrying out her plans to pay due homage to her dead husband she is condemning her daughters to a sterile and cheerless existence:

> During our eight years of mourning, no wind from the street will enter this house! Pretend we have sealed up the doors and windows with bricks. That was how it was in my father's house, and in my grandfather's house. (Lorca 205)

Like many of Lorca's works, *The House of Bernarda Alba* features strong women characters, and during the last half of the twentieth century the play has been given numerous stagings that emphasize its "feminist" themes and values. There is no doubt that both historical and personal factors combined to lead Lorca to make women central to his three tragedies of contemporary Spanish life (*Blood Wedding, Yerma,* and *The House of Bernarda Alba*): "His experiences abroad and the circumstances at home," writes Julianne Burton, "sensitized him to the social realities of the Spain of his day: an archaic, hypocritical, and crippling morality; a hierarchical, even tyrannical family structure; extreme social stratification and exploitation of the humbler sectors; and a social-sexual code which privileged men at the expense of women's autonomy, participation, and self-realization" (Burton 260). But there is ongoing debate as to the genuineness of Lorca's portraits of women, or, indeed, his motives in portraying them and men in terms of essential, gendered qualities and experiences. What Lorca takes to represent "masculinity" or "femininity," in other words, seems to be as much a cultural as biological question; as one recent commentator puts it, "there is no longer a Lorca without Foucault and Freud, without feminisms, theories of the body, without attendant questions about the bases and relationships between manifestations of heterosexuality and homosexuality, between various emerging eroticisms...and questions, not least, about the interrelatedness of romantic desire, sexuality, gender and social roles and conventions" (Perriam 151–52).

That the women in Bernarda's household are subject to an unseen but all-powerful masculinity is consistent with Lorca's general interest throughout his playwriting career in depicting tragic characters who fall victim to elemental forces—sexual, economic, cultural—above and beyond their ken or control. Hence in *Blood Wedding*, for example, the Bride expresses her misfortune as one who succumbs to the irresistible pull of *eros*: "I didn't want to! Your son was what I wanted, and I have not deceived him. But the arm of the other dragged me—like the surge of the sea, like a mule butting me with his head—and would have dragged me always, always, always! Even if I were old and all the sons of your son held me by the hair!" (Lorca 102)—hence, too, the peculiar claustrophobia that infects the staged world in *The House of Bernarda Alba*. The drama is structured in terms of a tension between onstage and offstage, between, on the one hand, the repressive, claustrophobic environment within Bernarda's house, and, on the other, the larger, seemingly less restricted world beyond the walls of the house that the audience cannot see but whose qualities and people are described in some detail by the different characters.

In contrast to the mimetic and exclusively "female" space inside the house of Bernarda Alba, the diegetic space of the play is fully peopled by men and images of masculinity as the women's conversation turns again and again to men—husbands, gypsies, priests, an old suitor, the image of a black man embroidered on a canvas bag. One could go so far as to say that in becoming the constant subject of the female characters' conversations, thoughts, and longings, in some respects the men of *The House of Bernarda Alba*, like Godot or Glycerium, acquire more histrionic authority—it is indeed a kind of "realness" of presence—than if they were to stand on stage fully visible to an audience. The audience is constantly made aware of events that take place outside the walls of the house: Poncia, Bernarda's maid, tells an exciting story about how a woman named Paca la Roseta was "kidnapped" more or less willingly the previous night by several men and taken to an olive grove for a nighttime of sex; a group of harvesters offstage sings lusty songs as they come in from working in the fields; and a breeding stallion is heard violently striking the walls of his stable in an attempt to get access to the mares in heat. And woven throughout the play like a leit motif is the figure of Pepe el Romano, a young man who is betrothed to Angustias but simultaneously carrying on an affair with Adela.

Lorca depicts these two spaces, one mimetic and the other diegetic, at least in part in terms of a repressive political surveillance. The play has often been interpreted as an indictment of fascism, where the condition of being perceived is to be without power or autonomy, while to be invisible carries with it a measure of authority and invulnerability. Thus the women in Bernarda's household jealously try to control one another through mutual surveillance—a state of being from which there is no relief, only ever-increasing visibility—even as their imaginations are held captive by events that take place beyond the walls of the house. What I would emphasize is that Lorca's dramatization of a masculine potency—for the play is surely that, at least in part—takes the particular form of obscuring it so as to stress its reality somewhere in the offstage as a product of the imagination. "Lorca," says Chris Perriam, " has his patriarchs, adulterers, rugged men of the soil and the horse...inhabit a hyper-signifying space, which leads them into iconic stasis; so much so, in *La Casa de Bernarda Alba*, that Pepe el Romano is so impossibly handsome, so much a virile force, that he has become a *deus ex machina* as well as the local stud" (Perriam 162).

A more recent play that brings what is offstage centrally into imaginative view is a work by the American playwright Maria Irene Fornes, *Fefu and Her Friends* (1977). *Fefu*, which is probably Fornes' best-known work, is set in a New England country estate during the spring of 1935; seven women have come to Fefu's house for a visit (it is not clear why), and the play documents their conversations and activities over the course of a single afternoon. They talk about love, about history, politics, and philosophy, but the subject always circles back to men and women and their relationships. Like the recent television comedy "Sex and the City," Fornes' play shows the extent to which the lives even of learned and articulate women are dominated by men and a culture that is built largely on masculine values, beliefs, and even masculine language. At one point one of the women, Emma, channels one of Shakespeare's sonnets ("Not from the stars do I my judgment pluck") in speaking to an effigy of Fefu she has constructed. But perhaps the most sinister instance of men's usurpation of women's identities involves

Julia, a strange and fragile character who seems to be alienated tragically from any genuine sense of self. Sitting in a wheelchair (a visible trope for her crippled spirit), Julia speaks a monologue in which she describes being tortured and brainwashed. It is not clear whether the events she describes really happened or whether she simply hallucinates them, but the degree of her self-alienation and self-loathing is clear in any case; she sits up *"as if pulled by an invisible force"* and mouths dully words she calls "my prayer":

> The human being is of the masculine gender. The human being is a boy as a child and grown up he is a man. Everything on earth is for the human being, which is man. To nourish him—There are evil things on earth, and noxious things. Evil and noxious things are on earth for man also. For him to fight with, and conquer and turn its evil into good. So that it too can nourish him—There are Evil Plants, Evil Animals, Evil Minerals, and Women are Evil—Woman is not a human being. She is: 1—A mystery. 2—Another species. 3—As yet undefined. 4—Unpredictable; therefore wicked and gentle and evil and good which is evil. (Fornes 35)

Fornes, like Glaspell, writes theatre from what has been called a woman's point of view (Sontag 9). Beginning her career as a writer associated with off-off Broadway theatre groups in the 1960s, Fornes has for more than forty years been a relentlessly experimental playwright. One of the ongoing topics of interest in *Fefu and Her Friends* is Fornes' innovative requirement that the four separate scenes of Part 2 are to be played simultaneously in different spaces, with all the scenes to be repeated until each member of the audience (which had been previously split into four groups) has viewed each different scene. Another interesting feature of *Fefu* is its cast, which consists exclusively of women, Fefu and her circle of friends. This is not to say the play is monastic; on the contrary, as was the case with *The House of Bernarda Alba*, men are a frequent, almost obsessive, topic of conversation among all the onstage characters. Fefu and her companions talk about men's natures, their strengths, and their relations with women. At times, the men's activities are even witnessed by women who stand at windows or at French doors and gaze through them to the "outdoors." These scenes are structured according to the classical method teichoscopy:

> Fefu:...We'll see. (Fefu *goes to the doors. She stands there briefly and speaks reflectively.*) I still like men better than women.—I envy them. I

like being like a man. Thinking like a man. Feeling like a man.—They are well together. Women are not. Look at them. They are checking the new grass mower.... Out in the fresh air and the sun, while we sit here in the dark. (Fornes, *Fefu and Her Friends* 13)

Scenes such as these are clearly intended to cause spectators to "see" what is taking place beyond their field of vision. But see how and for what purpose? Fornes surely had imaginative rather than mimetic perceptions in mind when she chose to represent the men only through the words and gestures of women; it seems, therefore, that she displaces the men to the offstage precisely so that they (i.e., masculinity or "maleness") can become realized in the form of spectators' own mental compositions. Because individual men cannot actually be seen they have to be "filled in" by spectators, who, drawing on their memories, quickly and unconsciously fill in particular images of individuals and then assume those details to be accurate, even though they have no substantive basis in the mimesis. The power of this kind of representation of men and/or masculinity depends on much the same dramaturgy of omission as Lorca's or Glaspell's virtual representations of unseen beings. Fornes evokes presence by naming an absence, and her portraiture depends on the kind of existential uncertainty that any such textual absence generates. The men Fefu describes exist only in words; they lack bodies altogether, and, apart from the few thoughts and interests that Fefu attributes to them, they have no ideas, emotions, or psychologies. We do visualize them, of course. When Fefu describes the men in the activity of "checking the new grass mower," an image of them readily enters the mind. One probably doesn't actually form a kind of mental video of individual men in motion, performing a sequence of actual physical movements—bending over, touching the handles, choke cable, or cylinder head fins. Rather one creates the sense of a visual experience of those acts, a "picture of picturing," so to speak. But we have so little actual visual evidence to rely on that the picture we create tends to be rather abstract—a simple schema or blueprint—onto which we can project the appropriate data acquired from the sensory areas of our brain.

The process of reifying absence seems particularly dependent in this case on the dematerialized object of the lawn mower. It would have been easy for Fornes to expand the foregoing scene along the lines of a conventional naturalistic production, to include a brief glimpse through the French doors of several men standing around a lawn mower. In this case the "new grass mower" would become a stage prop. But drawing attention instead as Fornes does to the object's absence, causes anyone who hears reference to it to want to picture how the lawn mower looks, making it into a kind of virtual prop, and it is this imagined solidity that gives depth and weight to the imaginary men offstage. In my own case, in the absence of any further description on Fornes' part of the machine and the grass, I summon up the memory of the terrace of my grandparents' house where I spent much of my childhood. I see in a wholly imaginary light a blood red power mower. I can call up the heft and substance of the machine as I tug on the hard rubber hand grips, and I experience in the back of my nostrils the delicate mix of metal, gasoline, and oil. I see that mower and that grass on the terrace above the orchard, but I alone see it. Everybody else who watches or reads Fornes' play sees something else, imagination in this case being dependent less on the stimulation of a rich verbal description (as in ecphrasis, for example) than on singular, private memories. Gilbert Ryle describes the same phenomenon from a philosophical point of view in *The Concept of Mind* (1950):

> Sometimes, when someone mentions a blacksmith's forge, I find myself instantaneously back in my childhood, visiting a local smithy. I can vividly "see" the glowing red horseshoe on the anvil, fairly vividly "hear" the hammer ringing on the shoe and less vividly "smell' the singed hoof. How should we describe this "smelling in the mind's nose"? (Gass 41)

Obviously a dramaturgy dependent to so great an extent on the unique memories of individual spectators allows for a relatively high degree of idiosyncratic and even contradictory responses. Presumably male spectators differ from female viewers in their imaginative reconstructions of the offstage figures; indeed, some people imagine the men of Fornes' play as representing omnipresent authority (Worthen, "Still Playing Games" 176), while others see them as secondary figures in a play dedicated exclusively to women (Moroff 36).

Yet the men of Fornes' play are as legitimately a part of the drama as are Glaspell's or Terence's women. Like Minnie Wright, Fefu's husband, the rest of the men, even the new grass mower, all are legitimately part of the play even though they never come on stage. Because the play is structured so as to invite spectators to construct their own images and then to assume (as is natural) that those images are accurate, men can be said to be important to the drama, present "in" it and central to it; it's not as if they've been banished from a club for bad behavior. Fornes has said that at first she wanted to have a male actor play one of the women's roles just so there would be at least one man on stage. "I guess I was afraid of having an all women cast," she said (Robinson

223). But audiences seem never to have thought of the play as lacking in men. On the contrary, according to Fornes, "[a]udiences at the American Place Theatre said in post-play discussions...that they felt his [Fefu's husband's] presence very strongly although he was never on the stage" (Robinson 223).

In all the dramas I have discussed so far in this chapter, imaginary vivacity comes about by way of a calculated erasure of enactment as that concept has been conventionally understood to be a necessary or even central component of drama. These works all share a common dynamic: each depends on a texture of allusions to one or more unseen characters but stops short of representing them explicitly, either by actual enactment or by verbal description. This principle is taken to its logical extreme in *Waiting for Godot*, where *not* to define Godot (let alone to embody him) is arguably the playwright's conscious artistic strategy. When asked who or what Godot was supposed to represent, Beckett reportedly told Alan Schneider that "if I knew I would have said so in the play." This may or may not be true (Beckett's answers to thematic questions about his work were often sibylline), but the impatience with which Beckett deflects Schneider's question suggests that by the time of their conversation he had surely come to understand that to represent Godot mimetically would be to diminish the character's hold on the imagination. Indeed Godot's openness to interpretation is essential to Beckett's dramaturgy: "it is irrelevant," says Scarry, "whether Godot is God, night, death, Pozzo, silence, or a war agent, for Beckett's interest is expectation or anticipation, that psychic state which defines one's present in relation to one's future. If Godot were identifiable, it would suggest that were Godot to come, man would stop waiting....In effect, waiting for Godot means 'waiting for Godot, Christmas, lover, summer *ad infinitum*' or 'waiting for the end of waiting'"(Scarry, *Resisting Representation* 94). I would add that not-seeing (Glycerium, or Minnie Wright, or any men whatsoever in Lorca's drama) is just as essential to artistic method, insofar as these plays all share with *Waiting for Godot* a common dynamic. Each depends upon a deliberate occlusion of the expected object in order to maximize that object's hold on the mind's "disocular eye" (Gass 41).

* * *

Perhaps the most rigorous application of diegetic techniques to theatre during the twentieth century belongs to Marguerite Duras. Duras aspires to a literarization of theatre reminiscent in some respects of neoclassical French tragedy. Like Racine's, Duras' theatre is primarily a recitative art form: her dramas are austere, restrained, and yet in some ways iconoclastic when compared with some well-known modernist paradigms for the theatre. By privileging the literary elements of drama (as opposed to the performative) she runs counter to one of the main avant-garde principles of theatrical art. Many influential twentieth-century theorists of modernism and post-modernism in the theatre, whether Adolphe Appia, the Futurist Thomaso Filippo Marinetti, Antonin Artaud, or, later, Richard Schechner, Julian Beck, and Judith Malina, advocated a non-referential stage in conjunction with a minimum of text. Even Gordon Craig, whose skepticism with regard to mimesis was at least as keen as that of Brecht (or even Plato), conceived of theatrical performance in terms of an autonomous, almost ritual purity. Theatre at its best was to be a triumph of pure mise-en-scène; a familiar objective of the avant-garde was to replace a literary, text-based theatre tradition with a "holy" theatre of presence.

The theatre of Duras, in contrast, consistently values the verbal over the visual, the literary over the performative; her playscripts and her film scenarios are biased consistently toward narrative. It might be objected here that Duras' work cannot help being emphatically "spectacular," insofar as much of her most important work has been in film. Yet even in cinema her practice has been to resist those formal elements of the medium having to do with conventional scenographic techniques for representing character and action. During a series of interviews with Xavièr Gauthier, for example (first published as *Les Parleuses* in 1974), Duras explained that the plot lines of her films sometimes were difficult to follow because they were missing certain sequences of shots:

> *M.D.*: Yes, I think that some shots are missing—it's not that they're missing, because often I film them and later I ignore them— what I call transition shots, the intermediary shots.
> *M.G.*: *That's it, they're cut; they're taken out.*
> *M.D.*: They're taken out, the shots that allow the viewer to move from one sequence to another. (Gauthier 62)

In crafting her films Duras makes use as well of some other principles of omission. She commonly rejects what has been called film's tyranny of the visual in favor of a cinematography where iconographic aesthetics are consistently undercut by phonetic structures. Much of her work in film has been characterized in terms of a tension between qualities that are literary, on the one hand, and pictorial, on the other. (In this respect her work clearly parallels Brecht's attempts to "literarize" the stage picture.) This tension is apparent in her earliest film *Hiroshima Mon Amour* (1959), a technically complex work developed as a collaboration between Alain Resnais, who directed the filming, and Duras, who wrote the screenplay. The film treats subjects that recur in all Duras' work: love, loss, and remembering. The action is set in Hiroshima, ten years after the end of World War II. A French actress has come to Hiroshima to work on a film about peace, and toward the end of her stay there she falls in love with a Japanese architect. Their brief affair brings to mind her painful memories of another forbidden love, a wartime romance in occupied Nevers with a German soldier. The two lovers had planned to flee to Bavaria to marry, but on the day they were to leave he is shot and killed by a French partisan, and she is publicly humiliated for associating with an enemy of France. Insane with grief, she hides for months in a cellar, then goes to Paris where, upon her arrival, the city is buzzing with the news of the bombing of Hiroshima. When the war ends she marries and has a family, keeping her wartime tragedy secret.

Hiroshima Mon Amour has been celebrated (although not universally so) for its novel mixing of sound and visuals. Whereas in conventional narrative films, sound had been used to support or to complement the visual story, in *Hiroshima* the two elements tend to work independently of one another so that sound often dampens what is visually present in order to evoke what is visually absent. This separation of elements can be identified, to take one example, in the different ways that music is used in the film. Sometimes the music that accompanies the visuals is naturalistic, as when during the peace parade one hears the sounds of Japanese children singing. But at other times the relationship between music and visuals is less determinate. During the introductory moments of the film, for example, the camera focuses alternatively on parts of bared bodies glistening with perspiration and on shots of hospital rooms, stairs, and dying patients. At the same time from somewhere in the background one hears the music from Giovanni Fusco's minimalist soundtrack: piccolo, flute, clarinet, and English horn. All these instrumental sounds work in dialectic with the foregoing visuals rather than as a score that is descriptive or explanatory of them. Meanwhile two voices not definitely identified with any of the images of men or women on screen are heard to carry on a conversation:

A man's voice, flat and calm, as if reciting, says:
He: You saw nothing in Hiroshima. Nothing.

(To be used as often as desired. A woman's voice, also flat, muffled, monot-
onous, the voice of someone reciting, replies:)

She: I saw *everything. Everything.* (Duras, *Hiroshima Mon Amour* 15)

Certainly *Hiroshima Mon Amour*, like most of Duras' film scenarios, reveals her impatience with cinematic conventions of realism, in partic- ular the "invisible" editing techniques and smooth camera styles that had been developed to help spectators buy into the illusion that they were watching reality unfold before their eyes. She is particularly con- cerned in her work to destabilize the conventional cinematic association between sound and visuals, wherein the aural components of a film— conversation, environmental noises, offscreen music—are used mainly to support the sequence of images. Duras' work for this reason is often associated with the self-reflexive films of "New Wave" directors such as François Truffaut, Jean-Luc Godard, and Resnais. Watching these films is in several important respects similar to watching a performance by Brecht's Berliner Ensemble, in that the films, like Brechtian perfor- mances, constantly call attention to their own material qualities and/or aesthetic figurations. New Wave films are films about filmmaking and film-watching (just as Brecht's *Lehrstücke* are plays about the process of making theatre), and they emphasize their own inherent "filmic-ness" rather than, say, their accuracy in representing visual reality.

Many of the same cinematic techniques Duras developed in con- junction with Resnais in *Hiroshima Mon Amour* are carried further in a subsequent film *India Song* (1975), where, as Lucy McNeece writes, Duras' "principal strategy...is to dethrone the image as the primary vehicle of narrative and heighten the use of offscreen sound" (McNeece 130). But this characteristic split between visual and verbal compo- nents is equally strong in her works for the stage. The short, haunting play *Savannah Bay*, for example, makes use of a stage divided into two separate theatrical spaces, one where the actors are seen and heard to speak, and one for the "set" wherein the events they try to remember are understood to have taken place:

> *An almost empty stage. In the foreground a table, with six chairs and two*
> *benches swathed in dust-sheets. Bare floor. All this occupies only a tenth of*
> *the total stage area, but it is here that* Savannah Bay *will be enacted.*
>   *Behind the foreground area and separate from it is a large set designed to*
> *suggest a vast empty landscape. A pair of curtains—wood painted to repre-*
> *sent red velvet—are parted to reveal a central vista stretching as far as the*
> *back wall of the theatre. This central space is flanked first by a pair of huge*

bright-yellow marble pillars rising right up to the roof, then by a lofty dark-green double door, flung open and resembling the door of a cathedral in the Po valley. Through the opening lies first a band of almost black light, then the sea. The sea, which reflects a changing light now cold, now scorching and now sombre, is framed, like the scroll of the Law.

    Thus the setting of Savannah Bay *is separate from the representation of* Savannah Bay*—uninhabitable by the women who are its protagonists, apart.* (Duras, *Four Plays* 98)

Like almost all of Duras' work, *Savannah Bay* is constructed as a memory play, and the "story" it enacts is a domestic tragedy of *eros* and *thanatos*. The characters include two women, an older woman named Madeleine, who is apparently an actress, and a young woman, not named, identified as Madeleine's granddaughter. These are the only characters who appear on stage, but two others who are absent from view are nevertheless "present" often in the two women's narratives: Madeleine's dead daughter (and the young woman's mother) and a man (the lover of the dead woman, also not named). Together, according to one of Duras' stage notes, "the two women summon up the memory of the girl who died in the warm sea of Savannah Bay" (Duras, *Four Plays* 112). Madeleine in particular seems driven to attempt to burrow inside the memory of her daughter who died. Her words have the quality of a text that has been repeated many times, as if the repetition were a sign of her struggle to come to terms with her loss. In trying to fathom her daughter's death (suicide? accident?) she mixes fact and fiction, confusing her own personal history with the various theatrical roles she has played. Over the course of the drama the memories and stories overlap to the extent that it becomes difficult if not impossible to know whether they come from the lives of three separate individuals or are mere fantasies of a single mind.

What is significant for Duras is not so much the persons or events about which the story is told, as the process of attempting to bring the material to light in the first place. *Savannah Bay* is structured as an attempt to remember painful events that might or might not be true, events that are in any case ultimately unknowable. The past the women try to recover—represented by the physical space that in a conventional theatrical production would constitute the mise-en-scène—is declared by fiat to be "uninhabitable" and "apart." The "Savannah Bay" of the title is impossible to locate; even the name is itself ambiguous. Multivalent and overdetermined, "Savannah Bay" is more than a physical space, whether real or allegorical: it refers, alternatively or simultaneously, to a person, a film in which Marguerite once acted, and a geography. This last identity is especially problematic. Ostensibly "Savannah Bay" has all the trappings of a real geographic location; it is described in the text as a swampy region at the mouth of the river Magra, which is, in fact, a real river in Tuscany flowing from the Apennines through the Magra Valley and into the Mediterranean Sea. But, of course, there is no "Savannah Bay" located anywhere on the Italian coast (though Duras may or may not have known of a real "Savannah Bay" situated in the British Virgin Islands). Thus "Savannah Bay," like the partly recognizable, partly fantastic settings of many of Duras' works, is endowed with conspicuously fictional qualities.

Even the ostensibly real environments specified in the stories, films, and plays of the so-called India Cycle are deliberately falsified, as Duras' notes for the published version of *India Song* specify:

> All references to physical, human or political geography are incorrect: You can't drive from Calcutta to the estuary of the Ganges in an afternoon. Nor to Nepal. The "Prince of Wales" hotel is not on an island in the Delta, but in Colombo. And New Delhi, not Calcutta, is the administrative capital of India. And so on. (Duras, *India Song* 5)

Simple visual imagery such as ecphrasis is sufficient if the playwright wishes an audience to imagine a set of characters and events taking place in a coherent extrascenic environment, whether real or fantastic. Pentheus's dismemberment, the battle of Salamis, even the "herring fleet" that Clov glimpses outside the closed space of *Engdame*, all these actions and locations can be relatively easily referenced in words that aid spectators in constructing virtual realities that have distinct pictural and sensory qualities. But there is something additionally and obstinately resistant to visual perception about Duras' dramaturgy, as if the author were intent on minimizing the place of mimesis in theatrical performance. "Duras treats vision as a verbal phenomenon," writes Susan D. Cohen; "to see means to imagine with words" (Cohen 93).

Brecht had as early as 1931 called for a "literarization" of theatre, but whereas Brecht was concerned mainly that dramatic performances be punctuated with narrative components such as titles, screens, or songs, Duras makes the opposition between diegesis and mimesis into an organizing formal principle. Her tendency to split off narrative modes of representation from more directly mimetic styles is present already in *Hiroshima Mon Amour*, which was once disparaged as too "literary"

(Georges Sadoul, quoted in Glassman 26), and it is at least implicit in the relatively late work for the stage *La Musica Deuxième* (1985), an expanded version of the earlier one-act piece *La Musica*. In this work, a man and a woman have come together to complete the final legal arrangements for their divorce. As they discuss various practical matters (such as what to do with their furniture), they talk about their marriage, and over the course of the play they realize they are still in love though each is now involved in a romantic relationship with someone else. The published text calls for a naturalistic production style, in terms of both the design of the set and the depiction of the characters and dialogue; their conversation all takes place in the lounge of a luxury hotel in northern France. But recently the work has been performed to favorable reviews a number of times as a staged reading with English subtitles projected behind the two readers, so that there is a kind of built-in distancing (or narrative) effect to the production that (at least when I saw it) was entirely consistent with Duras' generally mixed or mediated styles for theatre or film.

In the "India Cycle" of plays and films especially, this effacement of spectacle with language becomes almost a trademark. In *La Femme du Gange*, for example, the camera is trained relentlessly on a wind-swept beach, while characters independent of the scene recount the story of Lol Stein. A similar split between word and image is built into the film *India Song* (1975), a work originally commissioned as a play by Britain's National Theater and published first as a playscript (1973). The lengthy opening scene of the film shows an image of the setting sun to the accompaniment of a composition of offscreen music and voices. The camera records the sunset in one continuous shot, filmed in real time; it takes slightly more than three minutes for the sun's disk to sink below what appear to be a line of clouds on the horizon. Meanwhile we hear a female voice speaking and singing in a foreign language. The disembodied voice belongs to a character identified in the text only as the beggar woman. Never seen on camera, the beggar woman nevertheless plays an important role in both the stage work and the film. One senses her presence in the film, for example, not only when she is heard singing offscreen but also because of the hold she exercises on Duras' cinematic world.

In some ways the Beggar Woman enlarges the thematic dimension of the film: like the Gloucester subplot in *King Lear*, her story of displacement and loss mirrors and focuses the tragic fate of the central character of *India Song*, Anne Marie Stretter. We hear fragments of the beggar woman's story while the sun sets during the introductory minutes of the film, part of a conversation between two offscreen female voices:

Voice 1: A beggar woman.
Voice 2: Mad.
Voice 1: That's it.

The voices here function in some respects like the chorus in ancient Greek tragedy; they set the characters in context, and in their conversations they indulge in some broad interpretive speculations:

Voice 2: One day, she has been walking for ten years. One day, before her, the Ganges.
Voice 1: Yes. She stops.
Voice 2: That's it. All her children are dead, while she walks toward Bengal.
Voice 1: Yes. She leaves them, sells them, forgets them. Toward Bengal, becomes sterile.

But the chief formal purpose of this diegetic material is not thematic but formal, to undermine the power of the mimesis by creating an imaginative alternative to it. The music and words of the soundtrack call to mind an alternative story to the one that is played out on the screen, a story that is incomplete and out of reach. The beggar woman is neither seen nor understood—she speaks in Cambodian, a language presumably unintelligible except in its emotional registry to Western audiences for whom the film was made. Her words, like her person, are largely inaccessible, "not present," and her absence from the mimesis, therefore, creates a "blank" from which knowledge is understood to be lacking (Selous 137).

India Song, like the several other narratives and films belonging to Duras' "India Cycle" (including the novels *The Ravishing of Lol Stein*, *The Vice-Consul*, and *The Lover*, and the film *The Woman of the Ganges*), deals with the tragic love story of Anne Marie Stretter and Michael Richardson. The critical act in all of these works is Richardson's betrayal of his fiancé, Lola Valerie Stein, a few days before their wedding. The incident (most fully described in *The Ravishing of Lol Stein*) occurs during a ball held at a casino in a town on the coast of France, when Richardson becomes infatuated with an older woman, Anne Marie Stretter, whom he meets for the first time at the dance. Richardson no sooner sees her than he is hopelessly possessed by desire; he and Anne Marie Stretter spend the rest of the night together, dancing. When they finally leave the ball just before dawn, Lola, who has been watching them helplessly all evening, breaks down in hysterics.

The different works that comprise Duras' "India Cycle" each treat the same characters and events from different perspectives and in different artistic modes. In all of them, however, the scene of betrayal at the ball assumes primary importance in part because of its relative absence from the representation. Even in *The Ravishing of Lol Stein*, the scene is understood to have occurred ten years before the opening of the narrative; it is "represented" in that work only third-hand, recounted by a narrator who tells about the events of that night on the basis of a story he has heard from a person whom (he says) he no longer believes. Located in a distant and largely inaccessible past, this event, therefore, in the words of one critic, "circulates, deformed and deforming, through hybrid fragments of dialogue, gestures, elements of staging, and decor until it is no longer recognizable except as elegiac longing. Purified of its traumatic content, it functions abstractly as an origin, a haunting retrospective focus toward which all actions and feelings are magnetically drawn as to an exotic mystery" (McNeece 31).

India Song is a work that relies heavily on the elements of narrative as well as a provocative technical resistance to visual representation. The driving force in both film and play is a deliberate rejection of conventional mimetic spectacle; Duras shows only echoes or traces of people and events, never the things themselves, implying that representability itself is in question. In the film, after the introductory shot of the setting sun, the camera wanders slowly and without apparent purpose over an unidentified interior space; the décor is elegant, consistent with the furnishings of an ambassador's residence in India in the 1930s. At length people appear on screen, a man and a woman, dressed in evening wear, slowly dancing; they do not speak, and we receive information about them only indirectly, again by way of the same two women's voices heard earlier speaking offscreen. But these images themselves turn out to be mere reflections of people who remain unseen, out of the direct line of sight of the camera. The film is visually provocative because it is so emphatically organized as diegesis; the strategy is to enhance the quality of film-watching as a secondary or narrative experience rather than as a mimetic performance.

What made the writing of *India Song* possible, Duras has said, in her notes to the published stage version, was "the discovery, in *The Woman of the Ganges*, of the *means* of exploration, revelation: the voices external to the narrative" (Duras, *India Song* 6). Her insight, applied to a work for the stage, clearly documents an instance of mimesis being subjected to a mode of narrative discourse external to it. One pair (Voices 1 and 2) is feminine; the other two (Voices 3 and 4), says Duras, are men's voices. Both sets of voices speak from some extrascenic space about the characters who appear in view; they are at the same time within and without the work. We never know whose these voices are, or from what frames of reference they speak. They are understood somehow to be spectators to the story, but their most important characteristic is that their narrative voices are beyond embodiment, in every respect independent of the mimesis.

The two pairs of voices combine the filmic conventions of voice-overs and voice-offs, as those terms are commonly understood. ("Voice-over" typically represents the internalized speech of a character seen on-screen, as in an interior monologue or flashback, or the authoritative narrative voice heard in a documentary. "Voice-off," in contrast, usually refers to situations in which one hears a voice not belonging to a character visible on screen; it thus belongs technically to the diegesis.) Duras' voices are not in thrall to the mimetic dimension of either play or film: "they speak among themselves," Duras says, "and do not know they are being heard" (Duras, *India Song* 145). Because of that high degree of autonomy the pairs of voices have a weird authority: "There is always something uncanny about a voice which emanates from a source outside the frame," says Mary Anne Doane; "the narrative film exploits the marginal anxiety connected with the voice-off by incorporating its disturbing effects within the dramatic framework" (Doane 40–41). In one sense these voices are simple glosses on the action. Like the voices of the chorus in ancient Greek tragedy, they are alternately hesitant and passionate, sometimes wise, sometimes naive, yet always alive with an extraordinary sensitivity. But unlike Greek choruses the voices cannot be seen, and so they tend sometimes to complicate actions rather than clarify them. Often they are in tension with what is visible on stage. The technique is similar to that developed in *La Femme du Gange*, which Duras claims is actually two films, a "film of images" and a "film of voices." The relationship between the voices and the mimesis is disruptive rather than harmonious; the voices do not facilitate mimesis, and often they create obstacles to it.

We could think of the separation between visuals and sound track (or offstage voices) as a broken-up fabula in the manner of Brecht, but here the effects are mainly aesthetic and affective rather than political.

Partly the technique is reminiscent of some of the theoretical under-pinnings of theatrical modernism, especially as that "modernism" was characterized by an extreme prejudice against conventional realism and the illusion of a homogeneous décor. "Words catch the ear, plasticity—the eye," wrote Vsevolod Meyerhold; "The difference between the old theatre and the new is that in the new theatre speech and plasticity are each subordinated to their own separate rhythms and the two do not necessarily coincide" (Meyerhold 56). As some of Duras' narratives, because of their liberal use of direct discourse, tend to resemble un-mediated dramatic performances (Gross 402), so her plays and films, in turn, tend to be endowed with what seems to be a distinctly nar-rative quality. Scenic enactment is denatured by an inherently oral or narrative presence; one is struck by the generic indeterminacy of such works, an indeterminacy that is surely part of the author's conscious in-tent. Concerning the published text of *India Song*, for example, Duras calls the work simultaneously "play, text, film" (Grove Press transla-tion, 1976). Indeed both play and film share essential formal qualities and characteristics; each is shaped according to what Duras calls a self-contradictory dramatic form, one that figures central events according to what might be called an aesthetics of non-representability. "Duras," says McNeece, "will show us only echoes and traces of an event, never the thing itself, implying that events never exist in a unitary form that lends itself to mimetic spectacle" (McNeece 34).

Another method by which Duras undercuts the mimetic character-istics of her work is to displace the central event to a time and place beyond the mimesis. Duras writes that "the story of *India Song*...is in the past, legendary, a model" (Duras, *India Song* 10), which is to say that the work is conceived in the mode of narrative and not drama. For this reason the play, like the film, has the semblance of memory. Duras stipulates that what is seen in *India Song* is secondary, or derivative, while what is "real" is understood to be offstage. The moments in which people are seen to be doing things—engaging in conversation, dancing, or weeping—are always abstracted as parts of what is being told. The narrative voices consistently maintain a categorical separation between things seen and things heard; as for those scenes in the drama where the narrative voices are silent, Duras calls for a style of dialogue that achieves much the same distantiation: *"No conversation will take place on the stage,"* Duras writes, *"or be seen. It will never be the actors on the stage who are speaking"* (Duras, *India Song* 48). Of the overall theatrical effect, she says that *"[o]ne ought to get the impression of a reading, but one which is reported, that is, one which has been performed before.... To repeat: not a single word is uttered on the stage"* (Duras, *India Song* 50).

This rigid separation of the verbal and the visual can be understood as part of an interruptive diegesis with mimesis. The stage in *India Song* shows only what we know to have occurred already; when the play opens, Anne Marie Stretter is already dead. The way that *India Song* dramatizes her story from a narratological framework as an evaluated occurrence might be contrasted with conventional tragic dramas that make use of what might be called their destiny-generating function. Take Euripides' *Hippolytus* for purposes of comparison: as soon as that play opens, Aphrodite tells the audience that Hippolytus will not live to see another sun. This knowledge represents for us, the spectators, an exquisitely painful burden. But knowing what is in store for Hippolytus means that we understand his death differently from the way we would if it were something that we knew to have already occurred. Naturally one feels great sadness upon learning of so unfortunate and untimely a future event, but it is our knowledge of what *must* happen, of what is yet to come—and not our knowledge of what has happened—that makes Euripides' irony painful.

Much the same holds true also for those plays that proclaim their status as "history." Even though we know what is to happen in such dramas, even though the action is understood to have occurred at some point in the remote past, the events depicted are presented normally from within their own temporal frame rather than from without. That is to say, performances of plays such as *The Persians* or *Antony and Cleopatra* presuppose first that the space and time occupied by the actors correspond to the space and time represented in the fiction, and second, that this autonomy of the performance is not normally compromised by any external mediating system of communication, such as a film camera or extrascenic narrator, whether visible or not. "Drama," says Susanne Langer, "though it implies past actions...moves not toward the present, as narrative does, but toward something beyond....As literature creates a virtual past, drama creates a virtual future" (Langer 307).

In *India Song*, on the other hand, the audience is constantly being reminded of the tension between the "real" time of the unseen voices and the fictional time of the soundless events they see enacted before them, and this overlay of narrative accounts for the play's elegiac mode. The result is a kind of emptying out of mimesis. It is for this reason that Duras proscribes the representation of speech, even during those scenes when characters appear in situations in which one might expect their conversation to be the object of immediate perception. "*From the*

point of view of sound," Duras writes, "*the image, the stage plays the part
of an echo chamber*" (Duras, *India Song* 49). Were the characters to
speak, they would create a dramatic situation understood to represent
the "here and now," as it were, but as long as the characters and their
voices remain separated the images retain a narrative quality.

The voices coming from a place outside the mimesis convey a superior, expressly narratorial, sentience with respect to the story of Anne
Marie Stretter. The voices cast an entirely new light on the enactment; they have a kind of secondary imaginative effect. "The voices,"
says Duras, "have known or read of this love story long ago. Some of
them remember it better than others. But none of them remembers
it completely. And none of them has completely forgotten it" (Duras,
India Song 145). The voices come from a space that is beyond embodiment. They tend to deflect attention away from the physical actions
and even from the physical appearance of the characters. The voices call
attention to the fact that what we are seeing is "past" rather than "here
and now"; even as the play begins, when we see Anne Marie Stretter
and Michael Richardson displayed before us on a divan, Duras' stage
notes indicate that the scene is to be framed in terms of its historicity:

> *Voice* 2: After she died, he left India...
> *Silence.*
> *That was all said in one breath, as if recited slowly.*
> *So the woman in black, there in front of us, is dead.*
> *The light is now steady, somber.*
> *Silence everywhere.*
> *Near and far.*
> *The voices are full of pain.. Their memory, which was gone, is coming
> back.*
> *But they are as sweet, as gentle as before.*
> *Voice* 2: She's buried in the English cemetery...
> *Pause.*
> *Voice* 1: ...she died there?
> *Voice* 2: In the islands. (*Hesitates*) One night. Found dead. (Duras,
> *India Song* 14–15)

Just before the end of the drama Voice 4 tells of some of the last
moments of Anne Marie Stretter's life:

> The Young Attaché came back to the residency in the course of the
> night. He saw her.

She was lying on the path, resting on her elbow. He said: "She laid her arm out straight and leaned her head on it. The Vice-consul from Lahore was sitting ten yards away. They didn't speak to each other." (Duras, *India Song* 142–43)

As if in obedience to a script, these actions are repeated (Brecht would have said "demonstrated") by the actors:

What has just been related is what ANNE-MARIE STRETTER *does. She lays her face on her arm. Stays like that. The* VICE-CONSUL *looks at her, riveted to the distance between them.* (Duras, *India Song* 143)

The offstage (or offscreen) voices in *India Song* represent the mind of a narrator (or the implied author) whom we hear as well in the stage directions. They emanate from an offstage (or offscreen) space that reminds us constantly that the enactment we see is no more "present" than the absent reality registered by a photograph, and just as unrecoverable. Even some of the more naturalistic offstage sounds in the play contribute to this distancing effect. At one moment, for example, we hear the sound of rain: "*It is raining over Bengal. The rain cannot be seen. Only heard*" (Duras, *India Song* 19). Later we hear the cries of birds, "*so loud,*" Duras says, "*they are almost unbearable*" (Duras, *India Song* 127). Within a strictly mimetic dramaturgy, offstage sounds like these might tend to support the dramatic illusion, but in *India Song* the sounds of the rain and the birds seem created (as are the voices) to deprive the mimesis of its primacy. To the extent that the stage in *India Song* is perceived to be merely an "echo chamber," it tends to be "haunted," as Sharon Willis has said of *L'Eden cinéma* (1977), by the absent presence of a superior reality: "the other form of spectacle just beyond its boundaries, just behind the scene, just offstage" (Willis 11). Sound, not sight, and absence, not presence, are the essential qualities of Duras' theatrical experience.

* * *

The plays of the Austrian playwright and novelist Thomas Bernhard, like those of Duras, tend to be structured around an absence. Bernhard adopts a classical dramaturgy: his plays are tightly restricted in their depictions of time and space, and, typically, they use few characters. Also in the classical tradition, Bernhard's is a theatre less of action than recitation. His characters assemble on anniversaries, on birthdays, on homecomings or other ceremonial occasions; they speak lengthy monologues full of nostalgic reminiscences or bitter recriminations, and stage props sometimes include portraits or photographs of relatives or other memorabilia.

Violence and death pervade all of Bernhard's works for the stage. At the center of most of Bernhard's plays one finds some enormity, some act of violence that has been all but torn from conscious memory but nevertheless governs events. These events are never actually represented, and they are rarely directly referenced. But they nevertheless contaminate everything else. Past experiences of trauma, especially the experience of violent death, constitute a terrible absence and also an animating presence in the characters' lives. Their emotional geographies are determinate and yet in some respects unreal; even though background events seem to be laden with consequences, they remain unconfirmable because they are not fully available through representation. Such events seem to belong to some portentous but invisible dimension of history, yet even with the most heroic effort, they cannot be brought fully to light. Like the scene of Lol's betrayal at the ball in the Town Beach Casino, they belong to a unique class of things in art that evade the exclusionary logic of being either present or absent. They are part of what might be called the domain of the work of art: they are not "in" the text, and yet, like the peasant woman Heidegger stipulated was present in Van Gogh's painting of a pair of shoes, they are fully "there" nonetheless:

> From the dark opening of the worn insides of the shoes the toilsome tread of the worker stares forth. In the stiffly rugged heaviness of the shoes there is the accumulated tenacity of her slow trudge through the far-spreading and ever-uniform furrows of the field swept by a raw wind. On the leather lie the dampness and richness of the soil. Under the soles slides the loneliness of the field-path as evening falls. In the shoes vibrates the silent call of the earth, its quiet gift of the ripening grain and its unexplained self-refusal in the fallow desolation of the wintry field. (Heidegger 33–34)

It is the conspicuous absence of the defining or central event, its seeming resistance to being represented, that characterizes Bernhard's work for the stage. Death, the trauma of loss, historical atrocities, all function for Bernhard as generative impulses for theatre. In *Heroes' Square* (*Heldenplatz*, 1988), Bernhard's last work for the stage, shouts from the crowd that assembled in 1938 for the Nazis still echo in the imagination of Professor Schuster's wife. In *The President* (*Der Präsident*, 1975), the leader of a European nation spends a few days waiting for the assassination he knows will soon come. In *Eve of Retirement* (*Vor dem Ruhestand*, 1979), two former Nazis hold a party for Himmler's birthday, thirty years after his death; the highlight of the bizarre celebration is a nostalgic perusal of wartime photographs of soldiers and Holocaust victims. And events in *A Party for Boris* (*Ein Fest für Boris*, 1970) revolve around a mysterious accident that occurred some ten years earlier.

Bernhard's fixation on remembered pain may have its origins in part in personal history. Born an illegitimate child, he was raised largely by his maternal grandparents and during his late teens was committed to a sanitarium for what was considered an incurable lung disease. But his work is consistent as well with an existing commercialized culture that markets the experience of personal trauma much the way it does sneakers or handbags. The appearance of fraudulent memoirs (e.g., Binjamin Wilkomirski's *Fragments: Memories of a Wartime Childhood*, or, more recently, James Frey's *A Million Little Pieces*) not only is proof of the unscrupulousness of writers or the gullibility of a reading public hungry for harrowing "authentic" experience. More important, and more dismaying, it also suggests the existence of a kind of diffuse but widespread "memory envy," which can be satisfied only by grafting one's personal emotional biography onto larger, tragic structures of history. (Sylvia Plath was an early practitioner: in her famous poem "Daddy"— the contents of which are convincingly autobiographical but largely fantasized—Plath imagines her father as a Nazi and herself as a Jew in order to establish her identity as a "survivor.") In a social climate where victimhood can be used to claim the moral high ground—in this universal "soup of pain," as Ian Buruma put it recently (Buruma 30)—all truth, all history, is relative, and only personal feelings count as evidence. To the extent that ideas of individual (or collective) identity are based on one's exposure to events so overwhelming as to cause a deep and lasting psychic wound, the grounds for authenticity are ipso facto beyond all reach of comment or questioning.

The story of Caribaldi in *Force of Habit* (*Die Macht der Gewohnheit*, 1974) is especially instructive. Caribaldi is the manager of a small-time touring company of circus performers; his group includes a clown, a lion tamer (his nephew), a juggler, and his granddaughter, a high-wire walker. As a way to pass the time between rehearsals and performances, Caribaldi tries to lead his troupe through a performance of Schubert's "Trout Quintet." The idea initially was that playing music would help

Caribaldi and his troupe to relax, but ironically their practices only cause increased stress. They have attempted to play the "Trout Quintet" for more than twenty years, and not once in all that time have they succeeded in completing it. Always, something goes wrong: when the play opens at the beginning of yet another rehearsal, Caribaldi has lost the rosin for his bow, the Juggler is too petulant to play, the Clown keeps losing his cap, and each time he loses his cap he loses his place in the score. The Clown's folly makes the Granddaughter giggle, and then she loses her place also. Worst of all is the Lion Tamer. He is drunk, and his left arm is injured because one of his cats bit it; it is bandaged so heavily it looks like a club. All he can do is pound the piano noisily. The particular rehearsal that Bernhard dramatizes, like every other rehearsal for the past twenty years, never gets beyond the first few notes.

Caribaldi is a megalomaniac, and his petty ambitions are laughable. But even though *Force of Habit* is a very funny play, it is hardly a happy one. Habituated to sadism and violence, Bernhard's comedy has ominous political resonances: Caribaldi calls to mind the twentieth century's sad history of lunatics in power—Stalin, Hitler, Idi Amin, Pol Pot, Saddam Hussein. But the drama calls to mind other kinds of experience as well, psychic experience that is deeply painful and deeply private and also apparently incommunicable. Of Caribaldi's daughter, for example, we learn little but that

> One day
> she went up to the animals
> and the animals snapped at her
> bit right through the creature
> She was brave
> The doctors
> stitched her up well
> she was hardly stitched up
> when she took her fall. (Bernhard, *Die Macht der Gewohnheit* 77)

That Caribaldi speaks so seldom about the daughter is instructive; her physical absence from the play has as its counterpart her virtual presence in his behavior. The daughter's death is not shown; neither does Bernhard dramatize her father's immediate reaction to that event. Yet that woman can still be said to exist by way of her registration in Caribaldi's behavior and language. Mourning for her is etched into Caribaldi's mechanics of speech and body, as if his actions were in part attributable to the inability of a wounded organism to discharge or to disown the injuries to it. His actions immediately following her death, as they are narrated by the Clown, consistently reveal a pattern of distress and compensation. As is clear from the Clown's description, Caribaldi displayed no overt signs of grief or mourning; his only act was to commit his daughter's body to a hasty burial in an unmarked grave. This was followed by years of deliberate avoidance of the town in which the accident occurred. There was nothing else, save for a growing obsession on the part of Caribaldi to dominate the development of his daughter's daughter.

Of course, the Clown interprets this behavior as one more sign of Caribaldi's selfish, stony heart. But that opinion in turn arises from the Clown's belief that Caribaldi's personality and behavior are entirely under his control; he assumes that a person's actions are the visible manifestations of a coherent thinking, desiring being. But that may not be a fully accurate explanation of the "force of habit." Suppose one inquires why it is that Caribaldi now manifests an obsessive desire to force the granddaughter to follow the profession of the mother. One possible answer might be that in doing so, he is in effect compelling her to take on the body of the mother, filling the psychic and physical space that that woman once occupied. Under the circumstances, then, at the very least, it would be an error to call the mother "dead and gone." Someone from a culture not in the habit of valorizing autogenous notions of selfhood, for example, might indeed think they were seeing Caribaldi act as if he were trying to reanimate his daughter, who can be made to seem fully present in the body of the granddaughter.

Caribaldi's seeing the absent daughter in the person of the granddaughter has its counterpart in common experience of family relationships. We see how children invariably "repeat" their parents as genetic phenomena, for example: one often hears descriptions of a child as the "spitting image" of one or the other parent, as if the palpable presence of the parent was being effected before their eyes. Similar correspondences can manifest themselves as behavioral or temperamental phenomena, in that children almost invariably develop their individual contours of selfhood in terms of those patterns of affect and gesture displayed to them daily by their parents. (Wilshire 169)

Thus one could point to a Freudian "family romance" being dramatized in *Force of Habit*: Caribaldi's repeated attempts to dominate those around him have the peculiar cumulative effect of revealing his own powerlessness. He behaves like an automaton precisely because his entire life since the high-wire accident has been an unconscious striving to memorialize the missing daughter. His mechanistic actions are like those of the little boy Freud describes in "Beyond the Pleasure Principle," who, Freud theorized, repeatedly manipulated the appearance and disappearance of a small wooden spool in an effort to compensate the psyche for the grief he experienced when his mother was absent. In both instances, the compulsive behavior of an individual masks an unspoken and unacknowledged attempt to dominate a being who is beyond reach. In effect, then, the staged representation here becomes the outward sign of what itself cannot be represented. Bernhard affirms the existence of a mysterious "force of habit" by which the living are struck dumb in order that the dead may speak.

A similar structure of repetition that points beyond the stage can be found in the first scene of *A Party for Boris*. The play dramatizes one day in the life of a character identified only as "the Good One" (die Gute). This woman has no legs and moves about the stage on a wheelchair. She has for several years been married to a legless amputee, Boris (this is her second marriage), and each year on her husband's birthday she throws a raucous party for him and more than a dozen of his friends, themselves also legless. Over the course of an extended conversation about letter-writing with her servant, the Good Woman speaks a few times about her late first husband. The first reference is brief, a simple chronological marker:

> I have always torn my letters up thrown away
> in the whole ten years my husband is dead
> I have always torn up the letters.... (Bernhard, *Ein Fest für Boris* 16)

A subsequent reference to the first husband is more detailed; it concerns the circumstances of his death: "My husband was dead instantly / I was not dead / But I lost my legs" (Bernhard 24). Finally toward the end of the first scene, as she idly tries on different pairs of gloves, the Good Woman returns suddenly to the event itself:

> It wasn't as if I had been surprised by the accident
> it wasn't so
> (*taking the green glove off again*)
> To be dead
> to plunge into a light shaft
> to be dead like my husband
> In truth I haven't dreamed about him for weeks
> not for years.... (Bernhard 34)

Any one of these various statements is the end of a thread that leads back to an extratextual reality. The power of the representation here depends in large part on the special kind of uncertainty that textual absence generates because of a repeated return to the same subject. The first reference to the event seems an isolated and solipsistic image: *what* accident? we might say, just as we might say of Juliet's Nurse's speech, *what* earthquake? Yet this gap in the representation, this hole in the mimesis, as it were, is gradually made to seem an important part of the so-called fabula of the finished work. It provokes the flow of memory, gesture, and speech, to the extent that an absent but continually referenced event seems to direct the representation and to acquire the status of an enduring material presence. The accident constantly impinges on the Good Woman's consciousness; like an apparition, it extends its influence far beyond its relatively distant temporal location. Likewise the husband, dead for ten years, can be said to survive his death; he is forever pressing against the Good Woman, leaving marks on her consciousness and traces on her speech.

Within the conventions of a fully mimetic dramaturgy, the first husband is "dead" and cannot appear in the body of an actor in full view of the audience. But removing the husband from sight, like the decision to place the death of Agamemnon behind the closed doors of the palace, gives him (to recall George Steiner's description of messengers' speeches) "a paradoxical nearness and pressure." Never visible, he is in some sense always with her; his presence is a kind of ghosting, or scripting. Playwrights in earlier ages would have dramatized this situation by stipulating an apparition to appear or by placing on stage an actual image of the dead person, such as, for example, the ghosts of Darius (*The Persians*) or of Hamlet's father, or (within a naturalist dramaturgy) the portrait of General Gabler that hangs on the wall in Hedda's house. In terms of his effect on the represented action, however, the dead husband mentioned in Bernhard's play is no less influential a presence than these other more conventionally represented persons. This helps to explain why key relationships in Bernhard tend to be dramatized as forms of psychic cannibalism or in terms of an invasive "body snatching." "You were never truly together with one you loved," says the narrator of Bernhard's novel *Gargoyles*, "until the person in question was dead and actually inside you" (Bernhard, *Gargoyles* 17).

Yet another technique by which Bernhard makes absence present is the peculiar way that many of his characters quote, repeat, or recite the words of somebody else. Examples abound in both his fiction as well as his plays. The short novel *Playing Watten* takes the form of a single, rambling letter; this letter has purportedly been composed by a doctor to a man called Undt, who had requested the doctor to provide him with a complete written account of his *"perceptions, over a period of several hours, of the day before the day that you received this note"* (Bernhard, *Three Novellas* 64). The account sprawls over a great variety of subjects, from the physician's own personal troubles (his license to practice medicine has recently been revoked on account of his drug abuse), to traveling and tourism, to the sizes of buckles on his shoes. But the doctor always returns to the subject of the recent suicide of a man named Siller, who along with the doctor was part of a group of men who used to assemble regularly to play a card game called Watten. Much of the narrative is phrased conspicuously as reports, or even reports of reports, as in the following passage: "but when Siller was cut down from the tree, strangely enough, the traveler said, said the truck driver, the paper was no longer in Siller's pants pocket" (Bernhard 84).

Normally when narrators defer their accounts to the words of others, the technique can be understood in terms of its either tonal or thematic implications. Such deferrals help to bring about the kind of self-conscious narrative "distancing" one finds in some of the stories of Joseph Conrad, for example, as when the narrator of *Lord Jim* takes pains to situate his own tale within the more remote structure of Marlow's storytelling. Bernhard's characters' habits of quoting the words or ideas of others can be read also in terms of the characters' individual psychologies or politics. To defer repeatedly to the words of someone else can represent an abandoning of selfhood or political autonomy (Esslin 374; Honegger 64). There is indeed a political dimension to Bernhard's dramaturgy, a sense in which Bernhard's characters' incessant quoting is reminiscent of the way slogans circulate and are parroted mindlessly in contemporary political discourse: "tax and spend," "Bush lied," "liberal media," and so on.

But Bernhard's characters' repetitions can be seen also as a variant of the figure prosopoeia, in that Bernhard's dramaturgy enables persons otherwise absent from the mimesis to appear or to speak and to exercise influence on the representation. Bernhard does not put ghosts on stage, but he can achieve a remarkably similar effect of "ghosting" as various characters in plays such as *The President* and *Eve of Retirement* incessantly repeat what they've heard other people say. In nearly every case, no matter how stupid or banal the sentiment, no matter how ludicrous or immoral, the various characters simply pass on the words as if they were being channeled on behalf of the absent person.

This inscription of absence is effected in conventional theatrical performance, where—as Plato observed—it's customary for actors to pretend that the alien speech is their very own. But Bernhard foregrounds the practice, in effect making it a theatrical gesture at odds with its own theatricality. Thus in *The President*, the First Lady more than forty separate times defers to the words of a chaplain who never appears on stage: "Death completes / life / the chaplain says"; or, "Marcel Proust as the chaplain says / is a genius" (Bernhard, *The President* 38, 53). Meanwhile the President, trying to impress an actress with whom he's having an affair, pretends to the words of Voltaire: "Time my child / can't be turned back...Nothing lasts longer / than time / for it is the measure / of eternity" (Bernhard 81). In *Eve of Retirement*, this ghosting is even more widespread, as Vera and Rudolph recite the words of famous literati, ex-Nazis, their friends, and their parents. Vera defers repeatedly to her brother and her father: "Poverty is caused / by the poor themselves / Don't ever help the poor / father used to say" (Bernhard, *Eve of Retirement* 118). And Rudolph draws his words from Nazis, both imaginary as well as real ones, at one point speaking in the persona of Himmler: "You can keep your father's house he said / the gas plant won't be built where it was planned / I have given the order / to build it in a lot / one hundred and eighty kilometers away / from your father's house" (Bernhard 189).

As Bernhard configures it, prosopoeia has obvious formal effects; the recurring words and phrases ("father says"; "the chaplain says") function in part as a kind of unifying rhetorical element or as a leit motif in a musical composition. (Bernhard was, in fact, a trained musician.) But on Bernhard's stage the figure has important ethical consequences as well. That is to say, a large body of personal testimony regarding the effects of trauma suggests that in some ways the survivors of violence believe themselves to be guardians of vanished people or events; the manner in which Bernhard's characters constantly negotiate around the past suggests a perpetual reenactment of that past, a reenactment that constitutes a pattern of insistent, unappeasable mourning. As the Good Woman's broken legs permanently memorialize the history of her tragic encounter with the law of gravity, one could say that in marrying the cripple Boris she commemorates her dead first husband and indeed revivifies him. Likewise the two young boys in the early novella *Amras*, orphaned by the suicide of their parents, keep hearing their voices because for the children to "move on" from the traumatic event—as current pop psychology would phrase it—proves to be impossible:

> Our childhood, which was associated most intimately with our parents, precisely because we had never been shocked by them, but were always left to our own devices, not without their guidance, a very liberal and therefore exacting guidance...was present to us in those weeks as it had never been before...crazy in itself, it was a consolation to our craziness ...Often, averting our faces, we would be sitting facing each other in our catastrophic physical and mental condition, after long periods of cerebral turmoil, when suddenly my Walter would jump to the window, startled by a call...which, after a certain point in time, *I too* heard...but in the garden there was never even a hint of a person calling to us...yet for many weeks we always heard the call at the same time...quite clearly the voices of our parents calling. (Bernhard, *Three Novellas* 15)

This short novel, like so much of Bernhard's literary output, is obsessively concerned with acts of displacement and forgetting that are also acts of remembering; in his novels as well as his plays there exists an uncanny reciprocity between the characters depicted mimetically and those absent from the mimesis. Sometimes these characters clearly belong to a space that seems to have the qualities of a conventional offstage environment, as, for example, the dead first husband in *A Party for Boris*. But at other times characters or events are not so easily located within a coherent extrascenic space. Of the Nazis depicted or referenced in *Eve of Retirement*, Bernhard said in an interview published in *Der Spiegel* in 1980 that such people "are in me, just as they are in everyone else" (Calandra 149). I take it that Bernhard did not mean to express the commonplace sentiment that all humans carry within themselves a murderous potential. I think he spoke in a more literal sense: humans' memorializing of others reveals itself in ways that are various and elusive, manifest as much, sometimes, in their deepest inscriptions, as unconscious cannibalizing of habits of speech or gesture, or as acts of resistance or avoidance, rather than in the more familiar forms of remembered images of previous times and places.

In the spring of 2007 I taught *Eve of Retirement* to a class of Emory undergraduates; the day before the class, at 7:55 p.m. on April 30, Fox News concluded its nightly broadcast with the following two unrelated snippets of information: "Courtney Love says she's tired of wearing Kurt Cobain's pajamas to bed," and, "On this day in 1945, Adolf Hitler killed himself." The facts about Love's bedtime attire and Hitler's suicide were, of course, not presented as hard news; they were offered as trivia meant only to provide closure to the hour-long show. Yet listening to them, especially when taken together as a measure of the complex and ongoing life of a culture, the two infobits seemed to me to have an odd synchrony. The more I thought about them the more they seemed to frame exactly what I wanted to say to the students about Bernhard and his dramaturgy of absences and evasions, about the intricate and sometimes labyrinthine ways that the things we do and say can be laden with personal and political history. Love's remarks were published widely on Internet sites over the next several days, part of a story about her plans to auction some of Cobain's belongings. (Cobain, the lead singer of the rock group Nirvana and Love's husband, took his own life in 1994.) "I still wear his pajamas to bed," she said. "How am I ever going to form another relationship in my lifetime wearing Kurt's pajamas?" Love's words offer sad testimony to her crippling inability to come to terms with Cobain's death, to admit that he was beyond her reach; her sentient reanimation of Cobain each night as she puts on the pajamas of a dead man might easily have come from the pages of Thomas Bernhard (or, for that matter, from the work of Marguerite Duras), where, always, the offstage dead hold sway over the living and moving characters with a potency that is anachronous, relentless, and enfeebling.

As for what one might term corresponding memorializations of Hitler, those are (as might be expected) vastly more diffuse. As the Fox broadcast makes clear, the date of Hitler's death is literally still remarkable, important to remember in a way that the date of the deaths of, say, Albert Einstein, Winston Churchill, or Indira Gandhi are not. Or one could point to the ubiquitous Volkswagen Beetle, in continuous production from 1938 to 2003, and lately resurrected for the twenty-first century as the "New Beetle." For more than half a century the Bug's familiar body contours have remained essentially unchanged from a design for a "people's car" that the newly elected chancellor Adolf Hitler sketched and gave to Ferdinand Porsche in 1933. As for the less visible residues: in his recent study *Hitler and the Power of Aesthetics*, Frederic Spotts writes about the way the Führer even now haunts the West, more than half a century after his suicide in the bunker in Berlin:

> The fear is... evident in the way the art produced in the Third Reich has been hidden away or the public invited to mock it on the rare occasion

when some of it has been shown—a precise replay, mutatis mutandis, of the 1937 degenerate art exhibit. An exhibition of Hitler's paintings—or an exhibition of Hitler's and Churchill's paintings that would show vividly the difference in character of the two statesmen—is unthinkable. In an argument submitted in the spring of 2001 to the Federal Appeals Court in Washington, the United States Justice Department maintained that the very brush strokes of Hitler's watercolours have such incendiary potential that they must be guarded from the gaze of all but screened experts. Similarly, the governments of Germany and a number of other countries have decided that their citizens should not be permitted to read *Mein Kampf* and have banned it. Who is afraid of Adolf Hitler? Just about everyone. (Spotts 401)

Let me close this discussion of the aesthetics of absence with a brief reference to one more work for the theatre. In 1996 Brian Friel's play *Molly Sweeney* opened in New York. Friel writes about a woman named Molly Sweeney, who, blind from infancy, has her sight restored at age 41 after an operation. The operation is both a complete success and a complete failure: a success in the sense that her sight is restored, a failure, as she reminds us at length, because the world she now sees with her eyes is much less satisfactory than the one she saw with her imagination. Friel's play is obviously a commentary on the central premise of this book, that in some circumstances those things that are beyond the mimesis can be central to it, just as things that are not seen on stage may in some ways be experienced with greater intensity and satisfaction than if they were fully visible. But the theme of *Molly Sweeney* seems to me less important to this study than the play's formal design. Structured as a series of monologues—independent narratives, really—*Molly Sweeney* (in the words of one reviewer) is "less a play than a novella, a talking book." It is this aspect of Friel's play that I find most instructive; taken together with Brecht's creative experiments early in the last century with mixing showing and telling, it allows for modernist and postmodernist plays to be placed within a single pair of bookends. If as late in the twentieth century as 1996 playwrights were still striving to "literarize" theatre by making it less like drama and more like narrative—and if critics, so many years after Brecht, were still complaining about it—it may well be time to acknowledge that for twentieth-century playwrights, "mimesis through diegesis" is less a false start than a major aesthetic shift. Perhaps the displacement of enactment by narrative is evidence (as Puchner suggests) of a modern kind of "antitheatricalism," or perhaps it simply records a characteristic preference for structurally androgynous mixings of genres or for modes of theatrical performance that inhabit the margins between showing and telling. In any case the oscillation between narrative and enactment may eventually come to be seen as a marker of twentieth-century tastes and values in the art of theatre, just as naturalism was expressive of the late nineteenth century or neoclassicism of the Paris of Louis XIV.

WORKS CITED

Adorno, Theodor. "Commitment." Trans. Francis McDonagh. In *Aesthetics and Politics: Theodor Adorno, Walter Benjamin, Ernst Block, Bertolt Brecht, Georg Lukács*. London: Verso, 1997.

Aeschylus. *The Persians*. Trans. Seth G. Benardete. In *The Complete Greek Tragedies*. Ed. David Grene and Richard Lattimore. Chicago: University of Chicago Press, 1991.

Albee, Edward. *Zoo Story*. New York: Signet, 1959.

Aristophanes. *The Thesmophoriazusae*. Trans. Benjamin Bickley Rogers. In *Aristophanes*, Vol. III (Loeb Classical Library). Cambridge: Harvard University Press, 1963.

Aristotle, *Poetics*. Trans. S. H. Butcher. In *Criticism: The Major Texts*. Ed. Walter Jackson Bate. New York: Harcourt Brace Jovanovich, 1970, 19–39.

Barish, Jonas. *The Antitheatrical Prejudice*. Berkeley: University of California Press, 1981.

Barlow, S. A. *Euripides' "Trojan Women"*. Oxford: Aris and Phillips, 1986.

Barrett, James. *Staged Narrative: Poetics and the Messenger in Greek Tragedy*. Berkeley: University of California Press, 2002.

Barthes, Roland. *Camera Lucida: Reflections on Photography*. Trans. Richard Howard. New York: Hill and Wang, 1982.

Baugh, Christopher. "Brecht and Stage Design: The Bühnenbildner and the Bühnenbauer." In *The Cambridge Companion to Brecht*. Ed. Peter Thomson and Glendyr Sacks. Cambridge: Cambridge University Press, 1994.

Beckett, Samuel. *Waiting for Godot*. New York: Grove Press, 1954.

Benjamin, Walter. *Understanding Brecht*. Trans. Anna Bostock. London: NLB, 1973.

Bernhard, Thomas. *Ein Fest für Boris*. Frankfurt am Main: Suhrkamp Verlag, 1970.

———. *Gargoyles*. Trans. Richard and Clara Winston. New York: Vintage Books, 2006.

———. *Die Macht der Gewohnheit*. Frankfurt am Main: Suhrkamp Verlag, 1974.

———. *The President and Eve of Retirement*. Trans. Gitta Honegger. New York: Performing Arts Journal Publications, 1982.

———. *Three Novellas*. Trans. Peter Jansen and Kenneth J. Northcott. Chicago: University of Chicago Press, 2003.

Bevington, David, ed. *Medieval Drama*. Boston: Houghton Mifflin, 1975.

Blau, Herbert. "The Thin, Thin Crust and the Colophon of Doubt: The Audience in Brecht." *New Literary History* 21 (Autumn 1989): 175–97.

Bowers, Jane Palatini. "*They Watch Me as They Watch This*": *Gertrude Stein's Metadrama.* Philadelphia: University of Pennsylvania Press, 1991.

Brecht, Bertolt. *Brecht on Theatre: The Development of an Aesthetic.* Ed. and Trans. John Willett. Rpt. London: Methuen London Ltd., 1986.

_____. *The Measures Taken and Other Lehrstücke.* Trans. Carl R. Mueller. New York: Arcade, 2001.

_____. *The Messingkauf Dialogues.* Trans. John Willett. Rpt. London: Eyre Methuen Ltd., 1977.

_____. *Mother Courage.* Trans. John Willett. New York: Arcade, 1994.

Brenton, Howard. *Plays for the Poor Theatre.* London: Methuen, 1980.

Brockett, Oscar G., and Franklin J. Hildy. *History of the Theatre,* 9th ed. New York: Allyn and Bacon, 2003.

Burden, Michael. "Purcell's Operas on Craig's Stage: The Productions of the Purcell Operatic Society." *Early Music* (August 2004): 443–58.

Burke, Kenneth. *Counter-Statement,* 2nd ed. Los Angeles: Hermes, 1953.

Burton, Julianne. "The Greatest Punishment: Female and Male in Lorca's Tragedies." In *Women in Hispanic Literature: Icons and Fallen Idols.* Ed. Beth Miller. Los Angeles: University of California Press, 1983, 259–79.

Buruma, Ian. "The Joys and Perils of Victimhood." In *The Best American Essays 2000.* Ed. Alan Lightman. New York: Houghton Mifflin, 2000, 20–31.

Calandra, Denis. *New German Dramatists: A Study of Peter Handke, Franz Xaver Kroetz, Rainer Werner Fassbinder, Heiner Müller, Thomas Brasch, Thomas Bernhard, and Botho Strauss.* New York: Grove Press, 1983.

Carlson, Marvin. *Theories of the Theatre: A Historical and Critical Survey, From the Greeks to the Present.* Ithaca: Cornell University Press, 1993.

Case, Sue-Ellen. "Classic Drag: The Greek Creation of Female Parts." *Theatre Journal* 37 (October 1985): 317–27.

Chekhov, Anton. *The Cherry Orchard.* In *Anton Chekhov's Plays.* Trans. and Ed. Eugene K. Bristow. New York: Norton, 1977.

Chipp, Herschel B., ed. *Theories of Modern Art: A Source Book by Artists and Critics.* Berkeley: University of California Press, 1968.

Churchill, Caryl. *Mad Forest.* New York: Theatre Communications Group, 1996.

Cohen, Susan D. *Women and Discourse in the Fiction of Marguerite Duras: Love, Legends, Language.* Amherst: University of Massachusetts Press, 1993.

Craig, Edward Gordon. *The Mask: A Journal of the Art of the Theatre* (Florence, 1909–29). Reissued New York: Benjamin Blom, 1967.

Daileader, Celia. *Eroticism on the Renaissance Stage: Transcendence, Desire, and the Limits of the Visible.* Cambridge: Cambridge University Press, 1998.

Danto, Arthur C. *Analytical Philosophy of History.* Cambridge: Cambridge University Press, 1965.

de Jong, Irene J. F. *Narrative in Drama: The Art of the Euripidean Messenger-Speech.* Leiden: E. J. Brill, 1991.

Derrida, Jacques. *Writing and Difference.* Trans. Alan Bass. Chicago: University of Chicago Press, 1978.

Dissanayake, Ellen. *Homo Aestheticus: Where Art Comes From and Why.* New York: Free Press, 1992.

Doane, Mary Anne. "The Voice in the Cinema: The Articulation of Body and Space." *Yale French Studies* 60 (1980): 33–50.

Dryden, John. "An Essay of Dramatic Poesy." In *Criticism: The Major Texts*. Ed. W. J. Bate. New York: Harcourt Brace Jovanovich, 1970, 129–60.

Duras, Marguerite. *Four Plays*. Trans. Barbara Bray. London: Oberon Books, 1992.

_____. *India Song*. Trans. Barbara Bray. New York: Grove Press, 1976.

Durham, Leslie Atkins. *Staging Gertrude Stein: Absence, Culture, and the Landscape of American Alternative Theatre*. New York: Palgrave Macmillan, 2005.

Eisenstein, Sergei. *Film Form: Essays in Film Theory*. New York: Harcourt Brace, Meridian Books, 1975.

Elsom, John. *Erotic Theatre*. New York: Taplinger, 1974.

Esslin, Martin. "A Drama of Disease and Derision: The Plays of Thomas Bernhard." *Modern Drama* 23 (January 1981): 367–84.

Euripides. *The Bacchae*. Trans. William Arrowsmith. In *The Complete Greek Tragedies: Euripides V*. Ed. David Grene and Richard Lattimore. Chicago: University of Chicago Press, 1955.

_____. *Ion*. Trans. Ronald Frederick Willetts. In *The Complete Greek Tragedies: Euripides III*. Ed. David Grene and Richard Lattimore. Chicago: University of Chicago Press, 1955.

_____. *Medea*. Trans. Rex Warner. Ed. David Grene and Richard Lattimore. In *The Complete Greek Tragedies: Euripides I*. Chicago: University of Chicago Press, 1955.

Fleischman, Suzanne. *Tense and Narrativity: From Medieval Performance to Modern Fiction*. Austin: University of Texas Press, 1990.

Fludernik, Monika. *Towards a Natural Narratology*. New York: Routledge, 1996.

Fornes, Maria Irene. *Fefu and Her Friends*. New York: PAJ Publications, n.d.

_____. *Plays: Mud, The Danube, The Conduct of Life, Sarita*. "Preface," Susan Sontag. New York: Performing Arts Journal Publications, 1986.

Fountain, Henry. "Tarzan, Cheetah and the Contagious Yawn." *The New York Times*, August 24, 2004.

Freedberg, David. *The Power of Images: Studies in the History and Theory of Response*. Chicago: University of Chicago Press, 1989.

Friel, Brian. *The Faith Healer. Selected Plays*. Introduction Seamus Deane. London: Faber and Faber, 1984.

Frye, Northrop. *Anatomy of Criticism: Four Essays*. Princeton: Princeton University Press, 1957.

Gass, William H. *Fiction and the Figures of Life*. New York: Vintage Books, 1972.

Gauthier, Xavière, and Marguerite Duras. *Woman to Woman*. Trans. Katharine A. Jensen. Lincoln, NE: University of Nebraska Press, 1987.

Gems, Pam. *Three Plays: Piaf, Camille, Loving Women*. Middlesex, England: Penguin Books, 1985.

Gilbert, Daniel. *Stumbling on Happiness*. New York: Vintage Books, 2007.

Glaspell, Susan. *Trifles*. In *The Bedford Introduction to Drama*. Ed. Lee A. Jacobus. New York: Bedford/St. Martin's, 2005, 911–18.

Glassman, Deborah N. *Marguerite Duras: Fascinating Vision and Narrative Cure*. Rutherford, NJ: Associated University Presses, 1991.

Goldhill, Simon. "The Naive and Knowing Eye: Ecphrasis and the Culture of Viewing in the Hellenistic World." In *Art and Text in Ancient Greek Culture.* Ed. Simon Goldhill and Robin Osborne. Cambridge: Cambridge University Press, 1994, 197–223.

Goldwater, Robert. *Primitivism in Modern Art* (Enlarged Edition). Cambridge: Harvard University Press, 1986.

Gombrich, E. H. *Art and Illusion: A Study in the Psychology of Pictorial Representation.* Princeton: Princeton University Press, 1969.

Gross, Janice Berkowitz. "A Telling Side of Narration: Direct Discourse and French Women Writers." *The French Review* (1993): 401–11.

Gruber, William. "'Non-Aristotelian' Theater: Brecht's and Plato's Theories of Artistic Imitation." *Comparative Drama* 21 (1987): 199–213.

_____. "Building an Audience: Brecht's and Craig's Theories of Dramatic Performance." In *Essays on Twentieth-Century German Drama and Theater.* Ed. Helmut Hal Rennert. New York: Peter Lang, 2004, 71–78.

Gussow, Mel. "A Voice of His Own: Albee's Epiphany at 30." *The New York Times*, www.http://nytimes.com.books/99/08/15/specials/albee-gussow.html (accessed June 1, 2009).

Haber, Thomas Burns. "The Woman of Andros: Who Is She?" *The Classical Journal*, Vol. 50, no. 1 (1954): 35–39.

Hare, David. *The Blue Room: A Play in Ten Intimate Acts.* New York: Grove Press, 1998.

Hauptmann, Gerhart. *The Weavers.* Trans. Horst Frenz and Miles Waggoner. In *Masters of Modern Drama.* Ed. Haskell M. Block and Robert G. Shedd. New York: Random House, 1962, 128–56.

Haynes, Lewis L. Eyewitness to History, "The Sinking of the USS Indianapolis 1945," http://www.eyewitnesstohistory.com/indianapolis.htm (accessed May 28, 2009).

Heidegger, Martin. "Origin of the Work of Art." *Poetry, Language, Thought.* Trans. Albert Hofstadter. New York: Harper and Row, 1971.

Hills, Paul. *The Light of Early Italian Painting.* New Haven: Yale University Press, 1987.

Honegger, Gitta. "Acoustic Masks: Strategies of Language in the Theater of Canetti, Bernhard, and Handke." *Modern Austrian Literature 18* (1985): 57–66.

Horace. "Art of Poetry." Trans. W. J. Bate. In *Criticism: The Major Texts.* Ed. W. J. Bate. New York: Harcourt Brace Jovanovich, 1970, 51–58.

Hrotsvitha. *Dulcitius: The Martyrdom of the Holy Virgins Agape, Chionia, and Hirena.* Trans. K. M. Wilson. In *The Bedford Introduction to Drama*, 5th ed. Ed. Lee A. Jacobus. New York: Bedford/St. Martin's, 2005, 215–18.

Ingarden, Roman. *The Literary Work of Art.* Trans. George G. Grabowicz. Evanston: Northwestern University Press, 1973.

Isherwood, Christopher. "Existentialist Musings, Clinically Pondered in French." *The New York Times*, http://theater2.nytimes.com/2005/10/21/theater/reviews/21psych.html (accessed June 2, 2009).

Issacharoff, Michael. *Discourse as Performance.* Stanford: Stanford University Press, 1989.

Jameson, Fredric. *Brecht and Method.* London: Verso, 1998.

Jencks, Charles. *Post-Modernism: The New Classicism in Art and Architecture*. New York: Rizzoli International Publications, 1987.

Johnson, Samuel. *Preface to Shakespeare*. In *Criticism: The Major Texts*. Ed. W. J. Bate. New York: Harcourt Brace Jovanovich, 1970, 207–17.

Kane, Sarah. *Complete Plays*. London: Methuen, 2001.

Kermode, Frank. *The Sense of an Ending: Studies in the Theory of Fiction*. London: Oxford University Press, 1968.

Konečni, Vladimir, "Psychological Aspects of the Expression of Anger and Violence on the State." *Comparative Drama* 25 (1991): 215–41.

Kroetz, Franz Xaver. *Farmyard & Four Other Plays*. New York: Urizen Books, n.d.

Kubiak, Anthony. *Stages of Terror: Terrorism, Ideology, and Coercion as Theatre History*. Bloomington: Indiana University Press, 1991.

Langer, Suzanne. *Feeling and Form*. New York: Charles Scribner's Sons, 1953.

Lewis, Sian. "Barbers' Shops and Perfume Shops: 'Symposia without Wine.'" In *The Greek World*. Ed. Anton Powell. New York: Routledge, 1995, 432–41.

Lillo, George. *The London Merchant*. In *British Dramatists from Dryden to Sheridan*. Ed. George H. Nettleton, Arthur E. Case, and George Winchester Stone, Jr. New York: Houghton Mifflin, 1969, 601–23.

Lipking, Lawrence. "The Marginal Gloss: Notes and Asides on Poe, Valéry, 'The Ancient Mariner,' The Ordeal of the Margin, *Storiella as She Is Syung*, Versions of Leonardo, and the Plight of Modern Criticism." *Critical Inquiry* 3 (Summer 1977): 609–55.

Lorca, Federico García. *Three Plays: Blood Wedding, Yerma, The House of Bernarda Alba*. Trans. Michael Dewell and Carmen Zapata. New York: Farrar, Straus and Giroux, 1993.

Lott, Eric. *Love and Theft: Blackface Minstrelsy and the American Working Class*. New York: Oxford University Press, 1993.

Macintosh, Fiona. *Dying Acts: Death in Ancient Greek and Modern Irish Tragic Drama*. Cork: Cork University Press, 1994.

Marranca, Bonnie. "Introduction." In *Gertrude Stein: Last Operas and Plays*. Baltimore: Johns Hopkins University Press, 1995, vii–xxvii.

McAuley, Gay. *Space in Performance: Making Meaning in the Theatre*. Ann Arbor: University of Michigan Press, 1999.

McNeece, Lucy Stone. *Art and Politics in Duras' "India Cycle."* Gainesville, FL: University Press of Florida, 1996.

Meyerhold, Vsevolod. *Meyerhold on Theater*. Trans. Edward Braun. New York: Hill and Wang, 1969.

Middleton, Thomas. *Woman Beware Women*. In A. H. Gommes, Ed. *Jacobean Tragedies*. London: Oxford University Press, 1969, 307–98.

Mitgang, Herbert. "'Faith Healer' by Brian Friel Is Revived." *The New York Times*, www.http://theater2.nytimes.com/mem/theater/treview/html?res=9402E2D 91539F931A25752C1A965948260 (accessed June 1, 2009).

Moroff, Diane Lynn. *Fornes: Theater in the Present Tense*. Ann Arbor: University of Michigan Press, 1996.

Mueller, Roswitha. "Learning for a New Society: The *Lehrstück*." In *The Cambridge Companion to Brecht*. Ed. Peter Thomson and Glendyr Sacks. Cambridge: Cambridge University Press, 1994, 79–95.

Nagler, A. M. *A Source Book in Theatrical History*. New York: Dover, 1952.

O'Casey, Sean. *Three Plays: Juno and the Paycock, The Shadow of a Gunman, The Plough and the Stars.* New York: St. Martin's Press, 1981.

Olf, Julian. "The Man/Marionette Debate in Modern Theatre." *Educational Theatre Journal* 26 (1974): 488–94.

Olson, Elder. *Tragedy and the Theory of Drama.* Detroit: Wayne State University Press, 1966.

Ortega y Gasset, José. *The Dehumanization of Art and Other Writings on Art and Culture.* New York: Doubleday Anchor Books, n.d.

Parks, Suzan Lori. *Topdog/Underdog.* In *The Bedford Introduction to Drama*, 5th ed. Ed. Lee A. Jacobus. New York: Bedford/St. Martin's, 2005, 1706–1733.

Pavis, Patrice. *Dictionary of the Theatre: Terms, Concepts, and Analysis.* Trans. Christine Shantz. Toronto: University of Toronto Press, 1998.

Perniola, Mario. "Between Clothing and Nudity." In *Fragments for a History of the Human Body*, Vol. 2. Ed. Michel Feher. New York: Urzone, 1989, 236–65.

Perriam, Chris. "Gender and Sexuality." In *A Companion to Federico García Lorca.* Ed. Federico Bonaddio. Woodbridge, UK: Tamesis, 2007, 149–69.

Peters, Julie Stone. "Performing Obscene Modernism: Theatrical Censorship and the Making of Modern Drama." In Alan Ackerman and Martin Puchner, eds., *Against Theatre: Creative Destructions on the Modernist Stage.* New York: Palgrave, 2006, 206–30.

Pfister, Manfred. *The Theory and Analysis of Drama.* Trans. John Halliday. Rpt. Cambridge: Cambridge University Press, 1991.

Plato. *The Republic.* Trans. Paul Shorey. Cambridge: Harvard University Press (The Loeb Classical Library), 1978.

Plautus. *Casina.* Trans. Paul Nixon. In *Plautus: Vol. II.* Loeb Classical Library. Cambridge, MA: Harvard University Press, 1977.

Puchner, Martin. *Stage Fright: Modernism, Anti-Theatricality, and Drama.* Baltimore: Johns Hopkins University Press, 2002.

Rehm, Rush. *The Play of Space: Spatial Transformation in Greek Tragedy.* Princeton: Princeton University Press, 2002.

Robinson, Marc, ed. *The Theater of Maria Irene Fornes.* Baltimore: Johns Hopkins University Press, 1999.

Rose, Gillian. *Feminism and Geography: The Limits of Geographical Knowledge.* Minneapolis: University of Minnesota Press, 1993.

Ryono, Colin. "*Jaws* monologue," http://www.whysanity.net/monos/jaws.html (accessed June 1, 2009).

Scarry, Elaine. *Dreaming By the Book.* New York: Farrar, Straus, Giroux, 1999.

———. *Resisting Representation.* New York: Oxford University Press, 1994.

Schechner, Richard. "Foreword" to *Chinese Theories of Theater and Performance from Confucius to the Present.* Ed. and trans. Faye Chunfang Fei. Ann Arbor: University of Michigan Press, 1999.

Schnitzler, Arthur. *Reigen: Zehn Dialoge.* In *Arthur Schnitzler: Sein Leben, Sein Werk, Seine Zeit.* Frankfurt am Main: Fischer Verlag, 1981.

Scolnicov, Hanna. "Theatre Space, Theatrical Space and the Theatrical Space Without." In *The Theatrical Space*, ed. James Redmond. Cambridge: Cambridge University Press, 1987, 11–26.

Selous, Trista. *The Other Woman: Feminism and Femininity in the Work of Marguerite Duras.* New Haven: Yale University Press, 1988.

Senden, M. von. *Space and Sight: The Perception of Space and Shape in the Congenitally Blind before and after Operation.* Trans. Peter Heath. Glencoe, IL: Free Press, 1960.

Shakespeare, William. *The Complete Works.* Ed. Stephen Orgel and A. R. Braunmuller. New York: Penguin, 2002.

Shange, Ntozake. *spell #7: geechee jibara quik magic trance manual for technologically stressed third world people.* In *Modern and Contemporary Drama.* Ed. Miriam Gilbert, Carl H. Klaus, and Bradford S. Field, Jr. New York: St. Martin's Press, 1994, 608–23.

Smith, Anna Deavere. *Fires in the Mirror: Crown Heights, Brooklyn, and Other Identities.* New York: Anchor Books, 1993.

Smith, Susan Valeria Harris. *Masks in Modern Drama.* Berkeley: University of California Press, 1984.

Sontag, Susan. "Preface" to *Maria Irene Fornes: Plays.* New York: Performing Arts Journal Publications, 1985.

Sophocles. *The Women of Trachis.* Trans. Michael Jameson. In *The Complete Greek Tragedies: Sophocles II.* Ed. David Grene and Richard Lattimore. Chicago: University of Chicago Press, 1955.

Spotts, Frederic. *Hitler and the Power of Aesthetics.* New York: Overlook Press, 2004.

Stein, Gertrude. *Geography and Plays.* Mineola, NY: Dover Publications, 1999.

———. *Last Operas and Plays.* Baltimore: Johns Hopkins University Press, 1995.

Steiner, George. *Antigones.* New York: Oxford University Press, 1984.

Strindberg, August. *Miss Julie.* Trans. Elizabeth Sprigge. In *Masters of Modern Drama.* Ed. Haskell M. Block and Robert G. Shedd. New York: Random House, 1962, 94–111.

Taplin, Oliver. "The Pictorial Record." In *The Cambridge Companion to Greek Tragedy.* Ed. P. E. Easterling. Cambridge: Cambridge University Press, 1997, 69–90.

Terence. *Andria (The Girl from Andros).* Trans. Palmer Bovie. *The Complete Comedies of Terence.* Trans. Palmer Bovie, Douglass Parker, and Constance Carrier. New Brunswick, NJ: Rutgers University Press, 1974.

Walton, J. Michael. Ed., *Craig on Theatre.* London: Methuen, 1983.

Willett, John. *Brecht on Theatre: The Development of an Aesthetic.* New York: Hill and Wang, 1964.

Williams, Tennessee. *The Glass Menagerie.* New York: New Directions, 1970.

Willis, Sharon. *Writing on the Body.* Chicago: University of Illinois Press, 1977.

Wilshire, Bruce. *Role Playing and Identity: The Limits of Theater as Metaphor.* Bloomington: Indiana University Press, 1982.

Wilson, Edmund. *Axel's Castle: A Study in the Imaginative Literature of 1870–1930.* New York: Charles Scribner's Sons, 1931.

Worthen, William B. *Print and the Poetics of Modern Drama.* Cambridge: Cambridge University Press, 2005.

———. "Still Playing Games: Ideology and Performance in the Theater of Maria Irene Fornes. In *Feminine Focus.* Ed. Enoch Brater. New York: Oxford University Press, 1989, 167–85.

Wright, Elizabeth. *Postmodern Brecht: A Re-Presentation.* London: Routledge, 1989.

Yeats, William Butler. *Eleven Plays of William Butler Yeats*. Ed. A. Norman Jeffares. New York: Macmillan, 1964.

————. *At the Hawk's Well*. In *Masters of Modern Drama*. Ed. Haskell M. Block and Robert G. Shedd. New York: Random House, 1962, 429–32.

————. *Explorations*. Selected by Mrs. W. B. Yeats. London: Macmillan, 1962.

Zajonc, Arthur. *Catching the Light: The Entwined History of Light and Mind*. New York: Oxford University Press, 1993.

Zeitlin, Froma I. "The Artful Eye: Vision, Ecphrasis and Spectacle in Euripidean Theatre." In *Art and Text in Ancient Greek Culture*. Ed. Simon Goldhill and Robin Osborne. Cambridge: Cambridge University Press, 1994, 138–96.

Index

Printed in the United States
By Bookmasters